URBAN LAND USE

Community-Based Planning

URBAN LAND USE

Community-Based Planning

Edited by
Kim Etingoff

Apple Academic Press Inc. | Apple Academic Press Inc.
3333 Mistwell Crescent | 9 Spinnaker Way
Oakville, ON L6L 0A2 | Waretown, NJ 08758
Canada | USA

©2017 by Apple Academic Press, Inc.

First issued in paperback 2021

Exclusive worldwide distribution by CRC Press, a member of Taylor & Francis Group

No claim to original U.S. Government works

ISBN 13: 978-1-77-463689-3 (pbk)
ISBN 13: 978-1-77-188485-3 (hbk)

Library and Archives Canada Cataloguing in Publication

Urban land use : community-based planning / edited by Kim Etingoff.

Includes bibliographical references and index.
Issued in print and electronic formats.

ISBN 978-1-77188-485-3 (hardcover).--ISBN 978-1-77188-486-0 (pdf)

1. City planning--Case studies. 2. City planning--Health aspects--Case studies. 3. City planning--Citizen participation--Case studies. 4. City planning--Decision making--Case studies. 5. Land use, Urban--Planning--Case studies. I. Etingoff, Kim editor

| HT166.U68 2017 | 307.1'216 | C2016-906571-5 | C2016-906572-3 |

CIP data on file with US Library of Congress

Apple Academic Press also publishes its books in a variety of electronic formats. Some content that appears in print may not be available in electronic format. For information about Apple Academic Press products, visit our website at **www.appleacademicpress.com** and the CRC Press website at **www.crcpress.com**

About the Editor

Kim Etingoff

Kim Etingoff has Tufts University's terminal master's degree in Urban and Environmental Policy and Planning. Her recent experience includes researching with Initiative for a Competitive Inner City a report on food resiliency within the city of Boston. She worked in partnership with Dudley Street Neighborhood Initiative and Alternatives for Community and Environment to support a community food-planning process based in a Boston neighborhood, which was oriented toward creating a vehicle for community action around urban food issues, providing extensive background research to ground the resident-led planning process. She has worked in the Boston Mayor's Office of New Urban Mechanics, and has also coordinated and developed programs in urban agriculture and nutrition education. In addition, she has many years of experience researching, writing, and editing educational and academic books on environmental and food issues.

Contents

List of Contributors

Lindsay K. Campbell
US Forest Service, Northern Research Station, NYC Urban Field Station

Nick Cavill
Cavill Associates Ltd, 185A Moss Lane, Bramhall, Stockport, Cheshire SK7 1BA, UK.

Heejun Chang
Department of Geography, Portland State University

Alison K. Cohen
University of California Berkeley, School of Public Health, Division of Epidemiology, Berkeley, California, United States of America

Jason Corburn
University of California Berkeley, Department of City and Regional Planning & School of Public Health, Berkeley, California, United States of America

Carmen L. Cutter
Active Living Research, University of California, San Diego, USA; Department of Family and Preventive Medicine, 3900 5th Avenue, Suite 310, San Diego, CA 92119, USA.

Ding Ding
The University of Sydney, Sydney School of Public Health, Edward Ford Building (A27), Sydney, NSW 2006, Australia.

Brian Dyson
US Environmental Protection Agency

Jessa K. Engelberg
University of California San Diego/San Diego State University, Public Health Joint Doctoral Program, 3900 5th Avenue, Suite 310, San Diego, CA 92119, USA.

Klaus Gebel
James Cook University, Centre for Chronic Disease Prevention, PO Box 6811, Cairns, QLD 4870, Australia.

Andrew Geller
Office of Research and Development, U.S. Environmental Protection Agency

Maureen Gwinn
U.S. Environmental Protection Agency, National Center for Environmental Assessment

Robert W. Hoyer
Department of Geography, Portland State University

Scott Jacobs
US Environmental Protection Agency

Karima Kourtit
Department of Spatial Economics, VU University

Fran Kremer
U.S. Environmental Protection Agency

Michiel de Lange
New Media Studies, Utrecht University, Netherlands

Debbie Lou
Active Living Research, University of California, San Diego, USA; Department of Family and Preventive Medicine, 3900 5th Avenue, Suite 310, San Diego, CA 92119, USA.

Juliana A. Maantay
Lehman College, City University of New York

Melissa McCullough
Office of Research and Development, U.S. Environmental Protection Agency

Peter Nijkamp
Department of Spatial Economics, VU University

Mike Parker
Progress Health Partnerships Ltd, 5 Elmfield Road, Wigan WN1 5RG, UK.

James F. Sallis
Active Living Research, University of California, San Diego, USA; Department of Family and Preventive Medicine, 3900 5th Avenue, Suite 310, San Diego, CA 92119, USA.

Yan Dominic Searcy
Chicago State University, USA

William D. Shuster
US Environmental Protection Agency

Mya Sjogren
U.S. Environmental Protection Agency, National Center for Environmental Assessment

Michael Slimak
U.S. Environmental Protection Agency, National Center for Environmental Assessment

Elizabeth Smith
Office of Research and Development, U.S. Environmental Protection Agency

Chad Spoon
Active Living Research, University of California, San Diego, USA; Department of Family and Preventive Medicine, 3900 5th Avenue, Suite 310, San Diego, CA 92119, USA.

Tom Stockton
Neptune and Co., Inc.

Kevin Summers
National Health and Environmental Effects Research Laboratory, U.S. Environmental Protection Agency, Gulf Ecology Division

Erika S. Svendsen
US Forest Service, Northern Research Station, NYC Urban Field Station

Matthew Thompson
Department of Planning and Environmental Management, University of Manchester, Manchester, UK

Christina M. Thornton
University of California San Diego/San Diego State University, Public Health Joint Doctoral Program, 3900 5th Avenue, Suite 310, San Diego, CA 92119, USA.

Amanda L. Wilson
Active Living Research, University of California, San Diego, USA; Department of Family and Preventive Medicine, 3900 5th Avenue, Suite 310, San Diego, CA 92119, USA.

Martijn de Waal
Department of Media Studies, University of Amsterdam.

Acknowledgment and How to Cite

The editor and publisher thank each of the authors who contributed to this book. Many of the chapters in this book were previously published elsewhere. To cite the work contained in this book and to view the individual permissions, please refer to the citation at the beginning of each chapter. The editor carefully selected each chapter individually to provide a nuanced look at community-based urban land use planning.

The chapters included are broken into three sections, which describe the following topics:

- Chapter 1 uses Glasgow, Scotland, to investigate the relationship between land use decisions, resulting environmental conditions, and unequal health consequences for residents in different parts of the city.
- Chapter 2 is a literature survey that uncovers the substantial co-benefits of land designed for physical activity, including physical and mental health, social benefits, safety, sustainability, and economics.
- Chapter 3 proposes a set of urban health equity indicators to identify problems with the built environment and move cities toward better management of resources to create healthy communities.
- The authors of Chapter 4 look to the future in an attempt to describe how new media forms allow citizens to engage with and affect the built form of their communities.
- Chapter 5 identifies the successes and challenges of over a hundred community-based land stewardship groups in the northeastern United States.
- Chapter 6 focuses on the ways in which community organizations in low-income Chicago neighborhoods have been effective in working with city planning services that have few resources.
- In Chapter 7, the authors use a GIS-based collaborative decision tool to make land use decisions regarding vacant land redevelopment.
- Chapter 8 describes another community-input GIS model to project future models of land use along the urban fringe in Oregon, United States.
- Chapter 9 offers an interactive community planning process that incorporates multiple stakeholders with the goal of economically stimulating, conserving ecosystems, and meeting social needs in the port city of Amsterdam, Netherlands.

- In Chapter 10, community land trusts are presented as a way to democratically determine land use, in this case for the purpose of housing and community revitalization.
- Chapter 11 discusses a tool used by the U.S. Environmental Protection Agency's Sustainable and Healthy Communities Research Program to help communities plan for long-term economic, environmental, and social sustainability.

Introduction

Land use decisions are often an invisible part of urban communities across the globe. However, their effects are anything but invisible. Urban land use patterns directly impact residents along many axes, and do so unequally across segments of the population based on income and race. Fortunately, land use planners are increasingly recognizing the need for meaningful and skillful community engagement strategies in order to rectify the consequences of historical land use decisions, and to build healthier, stronger future communities through responsive land use planning.

This compendium covers a range of land use planning and community engagement issues. Part I explores the connections between land use decisions and consequences for urban residents, particularly in the areas of health and health equity. The chapters in Part II provide a closer look at community land use planning practice in several case studies. Part III offers several practical and innovative tools for integrating community decisions into land use planning. Taken as a whole, these chapters are a basis for furthering effective community input processes in urban planning. Together, planners and community members can make cities work better for all residents.

—*Kim Etingoff*

The population of Glasgow, Scotland has very poor health, compared to Scotland as a whole and the rest of the U.K., and even compared to other post-industrial cities with similar levels of deprivation and worklessness. Chapter 1 maps and analyzes several health indicators to examine health inequities within Glasgow and explore the spatial correspondence between areas of poor health, high deprivation, and proximity to derelict land, much of which is contaminated from past industrial uses. People in high deprivation areas are significantly more likely to be hospitalized for respiratory disease and cancer; have low birth weight infants; and for men to have much lower life expectancy than those not living in the high deprivation areas, indicating substantial health inequities within Glasgow. They are also much more likely to live in close proximity to derelict land. A methodology is described for creating an index (PARDLI–Priority Areas for Re-use of Derelict Land Index), combining scores for these health, deprivation, and environmental variables. The Index is used to select and prioritize communities for resource allocation and planning efforts, and is transferrable to other locations. Potential strategies are outlined for re-using the derelict land for the communities' public health benefit and neighborhood regeneration, including urban agriculture/community gardens, urban forestation, active and passive recreation areas, and linkage to existing open space networks and natural areas. This research is part of a larger project comparing Glasgow and New York City regarding the relationship between environmental health justice and aspects of the built environment.

To reverse the global epidemic of physical inactivity that is responsible for more than 5 million deaths per year, many groups recommend creating "activity-friendly environments." Such environments may have other benefits, beyond facilitating physical activity, but these potential co-benefits have not been well described. The purpose of chapter 2 is to explore a wide range of literature and conduct an initial summary of evidence on co-benefits of activity-friendly environments. An extensive but non-systematic review of scientific and "gray" literature was conducted. Five physical activity settings were defined: parks/open space/trails, urban design, transportation, schools, and workplaces/ buildings. Several evidence-based activity-friendly features were identified for each setting. Six potential outcomes/ co-benefits were searched: physical health, mental health, social benefits, safety/injury prevention, environmental sustainability, and economics. A total of 418 higher-quality findings were summarized. The overall summary indicated 22 of 30 setting by outcome combinations showed "strong" evidence of co-benefits. Each setting had strong evidence of at least three co-benefits, with only one occurrence of a net negative effect. All settings showed the potential to contribute to environmental sustainability and economic benefits. Specific environmental features with the strongest evidence of multiple co-benefits were park proximity, mixed land use, trees/greenery, accessibility and street connectivity, building design, and workplace physical activity policies/programs. The exploration revealed substantial evidence that designing community environments that make physical activity attractive and convenient is likely to produce additional important benefits. The extent of the evidence justifies systematic reviews and additional research to fill gaps.

As the urban population of the planet increases and puts new stressors on infrastructure and institutions and exacerbates economic and social inequalities, public health and other disciplines must find new ways to address urban health equity. Urban indicator processes focused on health equity can promote new modes of healthy urban governance, where the formal functions of government combine with science and social movements to define a healthy community and direct policy action. An inter-related set of urban health equity indicators that capture the social determinants of health, including community assets, and track policy decisions, can help inform efforts to promote greater urban health equity. Adaptive management, a strategy used globally by scientists, policy makers, and civil society groups to manage complex ecological resources, is a potential model for developing and implementing urban health equity indicators. Urban health equity indicators are lacking and needed within cities of both the global north and south, but universal sets of indicators may be less useful than context-specific measures accountable to local needs. In chapter 3, the authors briefly outline an approach for promoting greater urban health equity through the drafting and monitoring of indicators.

Over the last few years, the term 'smart cities' has gained traction in academic, industry, and policy debates about the deployment of new media technologies in urban settings. It is mostly used to describe and market technologies that make city infrastructures more efficient, and personalize the experience of the city. In chapter 4, the authors propose the notion of 'ownership' as a lens to take an alternative look at the role of urban new media in the city. With the notion of ownership the authors seek to investigate how digital media and culture allow citizens to engage with, organize around and act upon collective issues and engage in co-creating the social fabric and built form of the city. Taking ownership as the point of departure, the authors wish to broaden the debate about the role of new media technologies in urban design from an infrastructural to a social point of view, or from 'city management' to 'city making.'

Urban environmental stewardship activities are on the rise in cities throughout the Northeast. Groups participating in stewardship activities range in age, size, and geography and represent an increasingly complex and dynamic arrangement of civil society, government and business sectors. To better understand the structure, function and network of these community-based urban land managers, an assessment was conducted in 2004 by the research subcommittee of the Urban Ecology Collaborative. The goal of the assessment was to better understand the role of stewardship organizations engaged in urban ecology initiatives in selected major cities in the Northeastern U.S.: Boston, New Haven, New York City, Pittsburgh, Baltimore, and Washington, D.C. A total of 135 active organizations participated in this assessment. Findings include the discovery of a dynamic social network operating within cities, and a reserve of social capital and expertise that could be better utilized. Although often not the primary land owner, stewardship groups take an increasingly significant responsibility for a wide range of land use types including street and riparian corridors, vacant lots, public parks and gardens, green roofs, etc. Responsibilities include the delivery of public programs as well as daily maintenance and fundraising support. While most of the environmental stewardship organizations operate on staffs of zero or fewer than ten, with small cohorts of community volunteers, there is a significant difference in the total amount of program funding. Nearly all respondents agree that committed resources are scarce and insufficient with stewards relying upon and potentially competing for individual donations, local foundations, and municipal support. This makes it a challenge for the groups to grow beyond their current capacity and to develop long-term programs critical to resource management and education. It also fragments groups, making it difficult for planners and property owners to work in partnership with them. The organizational networks are self-contained and do not include business or even legal groups, which may point to a gap between stewardship and environmental justice organizations. Chapter 5 suggests that urban environmental stewardship combines land management with the desires of civil society, the private sector and government agencies.

Chapter 6 explores urban planning office and community influence on land-use decision making in two poverty-stricken but redeveloping neighborhood areas in Chicago. The Department of Planning and Development in this study had marginal impact on land-use decisions due to administrative limitations. Community influence is moderated by the degree to which low-income housing advocates can act directly as developers and produce housing units. The research findings indicate that land-use decisions intended to benefit the low-income resulted not from community-based political conflict but more so from community organization cooperation with political actors.

An integrated GIS-based, multi-attribute decision model deployed in a web-based platform is presented enabling an iterative, spatially explicit and collaborative analysis of relevant and available information for repurposing vacant land. The process incorporated traditional and novel aspects of decision science, beginning with an analysis of alternatives, building on this analysis with a workshop to elucidate opinions and concerns from key decision-makers relevant to the problem at hand, then expanded by extracting and compiling fundamental objectives from existing planning efforts and previously published long-term goals. The model was then constructed as an open-source, web-based software platform for use as a process for exploring, evaluating, comparing, and optimizing fundamental, strategic, and means objectives. The resulting beta model, MURL-CLE, is intended to allow all interested parties, from stakeholders to decision makers, to consider alternative options for reuse of vacant land in a neighborhood in Cleveland, OH and to do so in a deliberative, transparent, and defensible process. The beta model is intended to be a platform for growth as a decision science tool and to provide a reproducible mechanism for considering any complex decision that attempts to incorporate multiple competing objectives and to allow an iterative process, as opposed to a prescribed solution or ranking of alternatives, for community decision making. The aims of chapter 7 are twofold: 1) present the USEPA developed beta-version of the decision analysis tool Maximizing Utility for the Reuse of Land (MURL; www.clemurl.org), and 2) evaluate current Cleveland land use information in light of the SDM process, and suggest how it can be tailored to SDM for land reuse planning for a single neighborhood in Cleveland, specifically Slavic Village.

The authors of chapter 8 describe a future land cover scenario construction process developed under consultation with a group of stakeholders from our study area. They developed a simple geographic information system (GIS) method to modify a land cover dataset and then used qualitative data extracted from the stakeholder storyline to modify it. These identified variables related to the authors' study area's land use regulation system as the major driver in the placement of new urban growth on the landscape; and the accommodation of new population as the determinant of its growth rate. The outcome was a series of three scenario maps depicting a gradient of increased urbanization. The effort attempted to create a simple and transparent modeling framework that is easy to communicate. The incorporation of the regulatory context and rules and place-specific modeling for denser urban and sparse rural areas provide new insights of future land conversions. This relatively rapid mapping process provides useful information for spatial planning and projects for where and how much urban land will be present by the year 2050.

Port cities are historically important breeding places of civilization and wealth, and act as attractive high-quality and sustainable places to live and work. They are core places for sustainable development for the entire spatial system as a result of their dynamism, which has in recent years reinforced their position as magnets in a spatial-economic force field. To understand and exploit this potential, chapter 9 presents an analytical framework that links the opportunities provided by traditional port areas/cities to creative, resilient and sustainable urban development. Using evidence-based research, findings are presented from a case study by employing a stakeholder-based model—with interactive visual support tools as novel analysis methods—in a backcasting and forecasting exercise for sustainable development. The empirical study is carried out in and around the

NDSM-area, a former dockyard in Amsterdam, the Netherlands. Various future images were used—in an interactive assessment incorporating classes of important stakeholders—as strategic vehicles to identify important policy challenges, and to evaluate options for converting historical-cultural urban port landscapes into sustainable and creative hotspots, starting by reusing, recovering, and regenerating such areas. This approach helps to identify successful policy strategies, and to bring together different forms of expertise in order to resolve conflicts between the interests (or values) of a multiplicity of stakeholders, with a view to stimulating economic vitality in combination with meeting social needs and ensuring the conservation of eco-systems in redesigning old port areas. The results indicate that the interactive policy support tools developed for the case study are fit for purpose, and are instrumental in designing sustainable urban port areas.

Emerging in the cracks of the ownership model are alternatives to state/market provision of affordable housing and public/private-led regeneration of declining urban neighbourhoods, centred on commoning and collective dweller control. Chapter 10 explores how the community land trust model can become an effective institutional solution to urban decline in the context of private property relations. It explores a case study of a CLT campaign in Granby, a particularly deprived inner-city neighbourhood in Liverpool, England. The campaign seeks to collectively acquire empty homes under conditions of austerity, which have opened up the space for grassroots experimentation with guerrilla gardening, proving important for the campaign in gaining political trust and financial support. This chapter discusses the potential of the CLT model as a vehicle for democratic stewardship of place and unpacks the contradictions threatening to undermine its political legitimacy.

A sustainable world is one in which human needs are met equitably and without sacrificing the ability of future generations to meet their needs on environmental, economic, and social fronts. The United States (U.S.) Environmental Protection Agency's Sustainable and Healthy Communities Research Program aims to assist communities (large and small) to make decisions for their long term sustainability with respect to the three pillars of human well-being—environmental, economic and social—and are tempered in a way that ensures social equity, environmental justice and intergenerational equity. The primary tool being developed by the Sustainable and Healthy Communities (SHC) research program to enhance sustainable decision making is called TRIO (Total Resources Impacts and Outcomes). The conceptual development of this tool and the SHC program attributes are discussed in chapter 11.

PART I

Why Is Community-Based Planning Important?

CHAPTER 1

The Collapse of Place: Derelict Land, Deprivation, and Health Inequality in Glasgow, Scotland

Juliana A. Maantay

1.1 INTRODUCTION—THE COLLAPSE OF PLACE

"The decline in [community] health is the inevitable outcome of the collapse of place" (Fullilove and Fullilove 2000).

Glasgow is Scotland's most populous city, with nearly 600,000 people. It covers an area of 68 square miles, and is located along the north and south banks of the River Clyde in West Central Scotland. [Figure 1] Since World War II (WWII) it has become one of the quintessential examples of a post-industrial city whose fortunes suffered a sharp decline and many of whose peoples' lives epitomize the tragedies of the "dependency culture" of the modern welfare state, rapid deindustrialization, urban blight, multi-generational worklessness, hopelessness, and random violence, some of which was instigated by faulty policies at the national level (such as those pertaining to trade labor unions, deindustrialization, and rapid slum clearances). Although the city began to turn itself around economically in the 1980s, these negative perceptions and realities remain an influence on the health status of its residents.

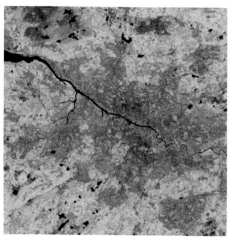

FIGURE 1.1 False-color satellite image of Glasgow, the purple areas showing the extent of urbanization. Data Source: NASA Socioeconomic Data and Applications Center (SEDAC), Urban Landsat: Cities from Space, (1999–2003)

The enduring poor health status of Glasgow's population has been welldocumented. According to the World Health Organization's Commission on Social Determinants of Health (2008), life expectancy for males in some Glasgow neighborhoods is only 54 years, a shocking figure for an affluent nation, especially one with universal access to health care. This life expectancy is lower than that of many less-developed countries whose people have minimal access to health care and are exposed to communicable diseases that by and large have been eradicated in Scotland. Although the alarming WHO statistics on Male Life Expectancy (MLE) only apply to a small part of Glasgow, overall MLE is still considerably lower than Scotland as a whole and the rest of the U.K. In comparison studies between Glasgow and other formerly highly-industrialized cities, both in the United Kingdom and abroad, Glasgow's mortality rates are much higher, and health is not improving as quickly (Halon et al. 2005; Mitchell et al. 2006; Taulbut et al. 2009; Walsh et al. 2010). In "*The Grim Reaper's Road Map,*" a recent atlas of mortality and other health indicators for all U.K. constituencies, Glasgow consistently shows up as a dark blotch on nearly every map, the dark color indicating the worst possible value for nearly every health variable being mapped (Shaw et al. 2008). Scotland as a whole suffers from inequalities in health, based on degree of deprivation, as compared to other parts of the U.K. and Europe, (MacIntyre 2007) but even within Scotland, Glasgow stands out as having worse health overall and sharper inequalities. There seems to be no obvious explanation of why Glaswegians tend to have such poor health outcomes, even when compared to cities such as Liverpool and Manchester, whose populations are similarly economically deprived, under stress from worklessness, and share a similar industrial history and culture (Gray 2008).

The extreme health disparities between Glasgow's population and the rest of Scotland and the U.K. include metrics such as low life expectancy, high proportion of low birth weight babies, and high rates of hospitalization for diseases such as diabetes, cancer, respiratory illness, and heart disease (Crawford et al. 2007; Glasgow Centre for Population Health 2008). However, these figures represent city-wide averages. In Glasgow, the figures for individual neighborhood health outcomes vary widely, and the health disparities/inequities that exist within Glasgow (differences in health outcomes amongst the wards, census districts, and neighborhoods of Glasgow) need to be comprehensively mapped and analyzed spatially in order to compare these within-Glasgow differences and ascertain the magnitude of health inequities within the city.

This paper reports on a mapping study of a selection of adverse health outcomes by census Data Zones within Glasgow, in relation to deprivation status and a specific category of environmental burdens—namely, vacant and derelict land (VDL). The main goal of this study is to develop an objective index, combining demographic, socio-economic, environmental, and health variables, which will be used to rank the neighborhoods in order of need, vulnerability, and exposure, thus providing decision-makers with guidance to prioritizing resources required to assist those areas most in need of mitigation.

1.1.1   The Deprivation-Health-Environment Connection

The connection between deprivation and poor health has been understood since at least the early 19th century (Chadwick 1842). After a fever epidemic struck Glasgow in 1843, Dr. Robert Perry, a surgeon at the Glasgow Royal Infirmary, mapped the homes of the fever victims in relation to socio-economic status in the various wards of the city, which also served as a proxy for housing and general environmental conditions. There was an extremely high degree of spatial correspondence between the two variables. On the map Dr. Perry produced, it is very striking how the affluent neighborhoods of the city had very few fever victims, but in the poorer areas - where clean water was not readily available, refuse and human waste piled up on the streets, overcrowding was rampant, housing did not have sanitary provisions, and industrial facilities with their attendant pollution were in close proximity to the homes -the epidemic was rampant (Perry

1844). Dr. Perry's maps and statistics successfully show the links between poverty, adverse health outcomes, and poor environmental conditions.

It may seem obvious to us today that there is a connection between deprivation and poor health, but there is also a link between deprivation and environmental burdens that has been less often acknowledged, especially prior to the environmental justice movement having brought it to the public's attention starting in the late 1980s - early 1990s (United Church of Christ's Commission for Racial Justice 1987; Bullard 1994; Johnston 1994; Bryant 1995). In the United States, this connection between high deprivation populations and proximity to environmental burdens has an important racial and ethnic component. However, because minorities are disproportionately represented in the lowest economic subgroups, race/ethnicity and socio-economic status are inextricably linked in the U.S., and it is difficult to separate the effects of income/poverty level from race (Maantay 2001; Maantay 2002). In Scotland, race/ethnicity may also be a factor in environmental health justice, but because of the relatively much lower numbers/proportions of racial/ethnic minorities in Glasgow, and the high numbers of poor non-minority people, it is believed that the multiple deprivation index alone suffices to measure the possible connections between health inequities and proximity to environmental burdens.

Reporting on a study pertaining to all of Scotland, Fairburn et al. (2004) state that "For industrial pollution, derelict land and river water quality there is a strong relationship with deprivation. People in the most deprived areas are far more likely to be living near to these sources of potential negative environmental impact than people in less deprived areas." Glasgow, having more of its share of industrial activities over the past few centuries than other parts of Scotland, would therefore also likely see these effects experienced more severely by the poorer populations than many other places.

There is another connection that has not been examined as much as would be expected – the link between poor health and environmental burdens. "Although much environmental justice research tacitly assumes that unequal environmental exposures produce geographic disparities in adverse health outcomes, very few empirical environmental justice studies have tested that assumption," (Grineski et al. 2013: 31).

This is a crucial part of the "triple jeopardy" of social, health, and environmental inequalities. "While the framework of 'environmental justice' has long been used to consider whether disadvantaged groups bear a disproportionate burden of environmental disamenities, perhaps surprisingly, the research fields of environmental justice and health inequalities have remained largely separate realms," (Pearce et al. 2010:522). In their 2010 study, Pearce et al. developed the Multiple Environmental Deprivation Index (MEDIx), for every census ward in the U.K., and found that multiple environmental deprivation increased as the degree of income deprivation rose, but more telling was that even after controlling for age, sex, and the socio-economic profile of each area, arealevel health progressively worsened as the multiple environmental deprivation increased. This points out "the importance of the physical environment in shaping health, and the need to consider the social and political processes that lead to income-deprived populations bearing a disproportionate burden of multiple environmental deprivation," (Pearce et al. 2010: 52).

1.1.2   The Landscape of Industrial to Post-Industrial Glasgow

In the 18th through the early 20thcentury, Glasgow was called "The Second City of the Empire," due to its importance as an industrial center and economic engine for the United Kingdom and the entire British Empire. Many of the industries prevalent in Glasgow and the Clyde River Valley at that time were "dirty" ones, with high levels of air pollution, toxic and dangerous chemicals routinely used in industrial processes, and environmental degradation of the surrounding areas. These industries included shipbuilding, steelmaking, coal mining, textile fabrication, dye works, brick works, rope works, tanneries, distilleries, railway locomotive works, cast iron foundries, chemical manufacturing, and the transportation industry (Hume and Moss 1977; Gibb 1983; Fraser and Maver 1996; Smyth 2000).

The population of Glasgow increased dramatically during the period of intensive industrialization, and by the end of the 19th and beginning of the 20th centuries, Glasgow had one of the highest population densities in the world: about 700,000 people concentrated in three square miles of central Glasgow. Most of the city's population lived in overcrowded conditions in 3- and 4-story sandstone tenement buildings, often entire large families in one or two rooms (Russell 1888; MacGregor 1967; Horsey 1990; Pacione 1995; Crawford et al. 2007; Faley 2008). There was little in the way of provisions for clean water and sanitation, and both communicable and chronic diseases were endemic. In the 1920s and 30s, these areas were acknowledged as being the worst slums in Great Britain. In the interwar and post-WWII periods, large portions of Glasgow's tenement neighborhoods were demolished and people were relocated to new housing estates and high-rise blocks of flats, often in peripheral areas at a distance from city centre. These new housing schemes, while offering modern amenities not previously available in the tenements (such as bathrooms within each dwelling unit, modern kitchens, and increased space and privacy), had important drawbacks: they were typically not well-constructed, were difficult and expensive to heat well, had inadequate transportation connections to the rest of the city, negligible shopping provisions, and helped to destroy the existing community life and strong social infrastructure of the old tenement neighborhoods (Horsey 1990).

By the 1960s, Glasgow was no longer the industrial powerhouse that it had been, owing to shifts in the global economy, changes produced by increasingly technological processes, and policy decisions at the national level designed to de-industrialize Scotland and diminish the strength of its highly unionized workforce. Factories and shipyards closed by the dozens, and the aftermath of this process was the visual blight of de-industrialization and abandonment in large swathes of the city, and the multi-generational worklessness that afflicts many Glaswegian families to this day. This problem of worklessness, in turn, has led to physical and mental health problems amongst the residents (Craig 2010).

There are many anecdotal (and somewhat controversial) explanations for Glaswegians' poor health, primarily circulating around individual behavioral issues of excessive drinking, smoking, drug use, violence, and poor diet (Craig 2010). Poor quality housing, with damp and mold, is also offered as a possible reason for poor health. The poverty and worklessness also translate into stress-related problems, mental breakdowns, feelings of hopelessness, loss of confidence in the future, alienation, and lack of control over their own lives, which can have direct and indirect physical health consequences (Burns 2012). Doubtless these are all valid reasons, and surely explain, at least partially, Glasgow's overall poor health, which is likely due to a complex combination of factors, and not any one thing.

But what about the external environment? Might not some of the high levels of poor health and health disparities be due to environmental factors? And even if causal links between environmental burdens and the overall poor health in Glasgow cannot be definitively demonstrated, wouldn't it be worthwhile to apply the precautionary principle in the absence of evidence to the contrary, and improve environmental conditions in the most deprived and least healthy places, where people are the most vulnerable? "When an activity raises threats of harm to human health or the environment, precautionary measures should be taken, even if some cause and effect relationships are not fully established scientifically," (Raffensperger and Tickner 1999). The precautionary principle implies that there is a social responsibility to protect the public from exposure to harm, when scientific investigation has found a plausible risk.

1.1.3 Environmental Justice and Vacant and Derelict Land in Glasgow

Environmental justice (EJ) is the concept that environmental benefits and pro-tections should be distributed equally amongst all populations, and environmental burdens should not disproportionately impact any subpopulation or region (Hofrichter 1993). Most often, however, low-income communities, immigrant neighborhoods, and communities of color bear a disproportionate burden of our pollution problems, whilst experiencing fewer environmental benefits and protection. There is a substantial body of evidence from previous research that has accumulated over the past 2 decades, as evaluated in a 2011comprehensive review of the literature by Brender, Maantay, and Chakraborty, that has found that proximity to environmental pollution is linked to poor health outcomes, and that this tends to disproportionately affect poor and minority populations (Brender et al. 2011).

The issues surrounding environmental justice in Scotland have been less often-researched, and EJ appears to be a more recent concern here than, for instance, in the United States. However, EJ in Scotland has been discussed in several books and papers covering environmental conditions, although health inequities are not the major focus, and differences amongst communities within any particular city are not addressed (Dunion 2003; Dunion and Scandrett 2003; Scandrett 2007; Scandrett 2010).

Due to the preponderance of vacant and derelict land in Glasgow, and the fact that the sites appear to be located primarily in the most deprived areas, this research concentrates on vacant and derelict land as an environmental burden and a potential environmental justice concern. There are 1,300 hectares of vacant and derelict land in Glasgow, with 927 individual sites, many of which are contaminated from past uses (Glasgow and Clyde Valley Strategic Development Planning Authority2010; Scottish Government 2012). This constitutes nearly 4% of Glasgow's total land area, and Glasgow's vacant/derelict land makes up over 12% of all Scotland's vacant/derelict land. Amongst Scottish Local Authorities, Glasgow has the highest amount of urban vacant land, in terms of both absolute number of hectares, as well as percentage-wise, by a very wide margin (Crawford et al. 2007). Approximately a third of the VDL sites have been vacant or derelict since 1990 or earlier, with about 10% of the VDL sites vacant or derelict since 1980 or earlier, or more than 30 years.

Over 60% of Glasgow City's population lives within 500 meters of a der-elict site, and over 92% live within 1,000 meters of a derelict site. [Figure 2]

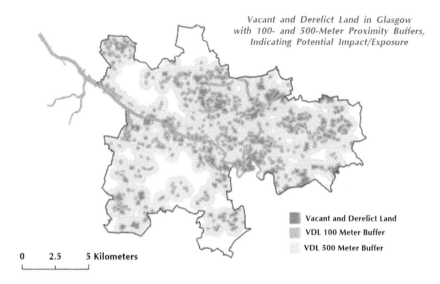

Vacant and Derelict Land in Glasgow with 100- and 500-Meter Proximity Buffers, Indicating Potential Impact/Exposure

Vacant and Derelict Land
VDL 100 Meter Buffer
VDL 500 Meter Buffer

0   2.5   5 Kilometers

FIGURE 1.2   Vacant and Derelict Land in Glasgow, with 100- and 500-meter proximity buffers, indicating areas of potential impact/exposures. Data Sources: U.K. Ordnance Survey (basemap layers); Vacant and Derelict Land Survey, Scottish Government, 2012 (VDL).

Five hundred meters is a generally accepted distance threshold in environmental analysis for assessing potential impact or exposure to contaminants, and was used in the Scottish Government's 2012 Vacant and Derelict Land Survey as an impact distance threshold (Chakraborty and Armstrong 1997; Neumann et al. 1998; Sheppard et al. 1999; New York City Mayor's Office of Environmental Coordination 2001; Scottish Government 2012). I have added a 100-meter buffer to the map to show a more conservative alternative exposure zone. A 1000-meter buffer, also used in the government's survey, would blanket virtually the entire city. Glasgow has the highest percentage of people living in close proximity to VDL of any local authority in Scotland (Scottish Government 2012). Most of this vacant and derelict land lies within the most deprived data zones - thus, VDL is an important and significant aspect of environmental injustice in Glasgow. The distribution of VDL disproportionately affects the poorest populations who, for many reasons, may also be the most vulnerable, health-wise. Previous research has demonstrated that given the same exposure to pollution or other environmental hazards, the people of lower socio-economic status will be more susceptible to their effects than the more affluent, due to existing health and quality-of-life vulnerabilities, material deprivation, and psychosocial stressors (O'Neill 2003). The impact of neighborhood blight and incivilities and the perceptual attitudes must also be taken into account in assessing risk to vulnerable populations (Ellaway et al. 2009). The effects of living near VDL transcend impacts on purely physical health, and encompass mental health as well.

What types of industries in the past were occupants of the now-derelict land, and what might the specific contaminants be that remain? Based on the known industries located in Glasgow (as mentioned in the section above) there are a number of carcinogenic or otherwise harmful substances likely in use and emitted to the environment. See the sidebar below for a list of potential contaminants.

BOX 1. POTENTIAL CONTAMINANTS LIKELY FOUND IN GLASGOW VDL

(Based on known former industries in Glasgow and likely land uses) (Data Source: United Kingdom DoE, 1995)

- Bonding agents, e.g., formaldehydes, and plastic compounds (polyurethane, acrylics and polyvinyl);
- Asbestos;
- Coal tar/creosote;
- Phenols;
- Cyanide & sulphur;
- Heavy metals, e.g., cadmium, lead, barium, chromium;
- Phytotoxic metals (damaging to plants), e.g., copper, nickel, zinc;
- Plating salts, e.g., various compounds, some containing cyanide;
- Aromatic & chlorinated hydrocarbons, e.g., benzene, trichloroethane, trichloroethylene, toluene, ethylene, xylene;
- Fuel additives, e.g. MTBE, hydrochloric acids, chloride & sulphide compounds;
- Solvents, e.g., kerosene, white spirit;
- Fuels and fuel byproducts, such as diesel, petroleum, aromatic hydrocarbon fractions, mineral oils, hydraulic fluid, engine oils, anti-freeze, petrol additives, diesel additives and detergents;
- Inorganic compounds, such as borates, bromide, fluoride phosphate & ammonium compounds (salts);
- Chlorinated organic compounds, e.g., TCE, PCE;
- Sizing agents, e.g., PVA, poly-acrylic acids.

These contaminants can live on in the environment long after the industry that produced or used them is gone. Exposure to the contaminants is an
on-going concern, particularly because children and youth often make use of
VDL for impromptu playgrounds and football fields. There are a number of
potential health impacts and pathways of exposure. Research has linked proximity to contaminated sites with pre-term and low birth weight infants, fetal
deaths, congenital malformations, heart disease, various cancers, and respiratory disease (Eizaguirre-Garcia 2000; Vinceti 2001; Litt and Burke 2002; Lit
et al. 2002; Baibergenova 2003; Malik 2004; Ding 2006; Kuehn et al. 2007;
Wang 2011).

Contaminant exposure may occur through airborne means, especially when
soil is disturbed; through dermal contact; or ingestion of soil or groundwater
(although Glasgow's drinking water supply does not rely on the City's groundwater). Aside from direct exposure to contaminants, VDL is often unsafe and
hazardous land to enter, and the effects of the resultant visual blight of vacant
land reduces quality-of-life and may result in additional day-to-day stressors
on residents (Greenberg et al. 1998), which can have direct effects on physical
health as well as adverse mental/emotional ramifications.

Many areas of Glasgow have become little more than "sacrifice zones"—
areas where the physical conditions are so poor that in an urban planning triage
situation, given limited resources, some planners and economists consider that
the sensible thing to do is to put the resources where there seems to be a hope of
a turnaround, rather than throwing good money after bad, as it were [personal
communications with Scottish urban planners and economists]. Thus, the very
worst areas in terms of deprivation and health frequently do not get the additional resources to make a difference, due to planners feeling helpless to effect
change in the face of a continuing downward spiral. This is not due to any mali-
cious intent or negligence on their part, but rather due to a sense of realism and
pragmatism about the limits of available resources. Another related problem,
which is common in virtually all cities and is not particular to Glasgow alone, is
that governmental resource allocation is often not based on a rational objective
assessment of need, but is decided on a more case-by-case basis, often driven by
political expediency, or from opportunities that arise unpredictably for private
investment. This analysis seeks to replace the subjective approach by providing
decision-makers with a more quantitative, evidence-based foundation for determining priority areas.

1.1.4 *Objectives*

"Rebuilding brownfields neighborhoods through an integrative public health and planning approach will be essential for improving the odds for sustainable redevelopment and securing long-term gains in public health," (Litt et al. 2002).

Taking into account the spatial distribution of deprivation and health inequities, and examining the spatial correlation between these indicators and the locations of VDL and potentially contaminated sites, where might we prioritize community participatory interventions to utilize these derelict lands for the benefit of the affected communities? In other words, based on the spatial analysis, which areas in Glasgow have high deprivation, poor health outcomes, large amounts of vacant and derelict land, and would benefit from additional neighborhood parks, natural areas, greenspaces, or other community uses?

What kinds of "ecological services" might these derelict lands provide the affected communities and the larger region? These ecological services might be flood control, stormwater management, urban agriculture, open space, natural areas, or recreational space for the surrounding communities or wider region. Temporary uses could also be considered, such as containerized gardening, or planting for phyto-remediation or phyto-stabilization, with eventual harvesting of the trees (Brack 2002; Eadha Enterprises 2012).

The aim of this research is not to prove causality between vacant and derelict land and adverse health outcomes. This would be extremely difficult to do, considering the lack of data available on three key variables: specifics on the actual type and magnitude of site contamination in Glasgow; records of individual health outcomes; and residential mobility (since many diseases, particularly cancers, have long latency periods). However, there is very likely to be a risk associated with living near many of these vacant and derelict sites, given the history of industrial land use in Glasgow, especially since even a vacant site formerly used for housing might have originally been built on land contaminated by industry.

Regardless of whether or not the actual risk of exposure involved can be demonstrated, the populations in these areas are very vulnerable on a number of levels: they are already suffering from higher than expected rates of many diseases, do not enjoy long life expectancy, and have to bear the stress of poverty and other forms of deprivation. Therefore, there is a strong environmental health justice imperative in determining which neighborhoods in Glasgow have the highest need for planning and implementation interventions and resource allocation.

1.2   METHODOLOGY AND ANALYSIS

1.2.1   Rationale for Decisions about the Spatial Analysis

The boundaries of the City of Glasgow were selected as the extent of the study area, in order to provide a high level of consistency, availability, and comparability of data. Some of this consistency would have been sacrificed if the study area extent had been expanded to include surrounding suburbs, which make up the Greater Glasgow region, and lie within several different local authorities. The government's census Data Zone (DZ) was selected as the most appropriate unit of data aggregation, since each data zone comprises ~750 people on average, which is small enough to get a reasonably fine-grained perspective of the issues, and to conduct detailed spatial analysis, but not so small that the statistical problem of "small numbers" would be a problem in most cases. There are 694 Data Zones in Glasgow. For some variables (e.g., Life Expectancy), it was necessary to use the larger-extent Intermediate Data Zone (IDZ) as the spatial unit, since data at a smaller extent would be unreliable due to small numbers, and for some variables the IDZ was the smallest extent available. There are 134 Intermediate Data Zones in Glasgow.

1.2.2   Exploratory Spatial Data Analysis

Exploratory Spatial Data Analysis (ESDA) is intended to allow relationships, patterns, and correlations to be revealed, clarified, and better understood. It is primarily used to generate hypotheses, and as a screening technique to indicate potential areas of fruitful further inquiry and research. A number of variables were mapped in order to formulate research questions and hypotheses, to investigate the issues on a first-pass screening basis, and to ascertain by visual inspection whether or not there are likely health inequity issues within Glasgow City. These variables are the following: the Scottish Index of Multiple Deprivation (SIMD); Health Decile, per DZ (derived from the SIMD's Health Domain); Rates of cancer hospitalization per 100,000, per DZ (CANCER); Rates of respiratory disease hospitalization per 100,000, per DZ (RESP); Low Birth Weight Infants as a percentage of all live births, per DZ (LBW); and Male Life Expectancy, per IDZ (MLE). These data were all obtained through the Scottish Government's National Statistics – Scottish Neighborhood Statistics website (Scottish Government 2009).

The SIMD is a weighted index, and is frequently used as a proxy metric for allover deprivation. It is compiled from 38 indicators in 7 domains, which are income, employment, education, housing, health, crime, and geographic access. The income and employment domains carry the most weight in the Index, at 56% combined. By contrast, the health domain, itself made up of 7 variables, is weighted at only 14%. The health indicators in the SIMD are: standardized mortality ratios; alcohol-related hospital episodes; drug-related hospital episodes; comparative illness factor; emergency admissions to the hospital; proportion of the population being prescribed drugs for depression, anxiety, etc.; proportion of low birth weight infants (Scottish Government 2009). The Scottish government uses a threshold of the most deprived 15% of the DZs for analysis and comparison purposes, particularly in longitudinal studies, looking at change over time. The SIMD was used in this analysis rather than a simple "income" or "poverty" variable, since the SIMD encapsulates a variety of deprivation measures, not just monetary ones. However, by mapping income deciles it was seen that there is almost total spatial correspondence between the worst income deciles and the 15% worst ranks (i.e., the highest deprivation levels) of the SIMD, by DZ. The SIMD rank data are for 2009, the most current year available at the time of the study. [Figure 3]

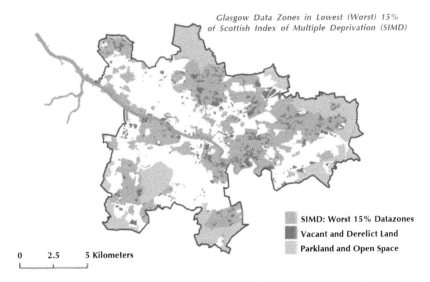

FIGURE 1.3   Glasgow Data Zones in Lowest (Worst) 15% of SIMD. Data Sources: U.K. Ordnance Survey (basemap layers); Vacant and Derelict Land Survey, Scottish Government, 2012 (VDL data); Scottish Index of Multiple Deprivation, General Report and Technical Report, Scottish Government Census, 2009 (SIMD data).

Health data are from 2010, covering the years 2005-2009. The health variables used in this study were selected as being salient factors in the overall poor health in Glasgow, and are fairly representative of the major types of health concerns. Arguments could be made that using more or different categories of health outcomes would have yielded better or different results. However, through consultation with a number of public health and medical geography professionals working in Glasgow and very familiar with its health conditions, it is believed that these variables as selected accurately capture the overall health status of each DZ. Additionally, because the SIMD's health domain for the most part includes different variables than the ones selected here, there was a lesser risk of magnifying or double-counting the effects of health, or having confounding factors, yet by using both indices, achieve good coverage of a variety of health outcomes. [Figure 4]

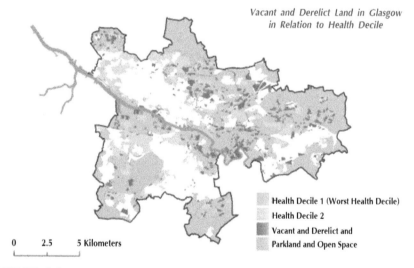

Vacant and Derelict Land in Glasgow in Relation to Health Decile

Health Decile 1 (Worst Health Decile)
Health Decile 2
Vacant and Derelict and
Parkland and Open Space

0   2.5   5 Kilometers

FIGURE 1.4   Vacant and Derelict Land in Relation to Health Decile. Data Sources: U.K. Ordnance Survey (basemap layers); Vacant and Derelict Land Survey, Scottish Government, 2012 (VDL data); Scottish Neighbourhood Statistics, Scottish Government, 2010 (health data).

These SIMD and health indicators were then examined in relationship to the location of the vacant and derelict sites, which were then also buffered with 500-meter and 100-meter exposure buffers. The vacant and derelict land data (non-spatial attribute data) were obtained from the Scottish Government's Survey on Vacant and Derelict Land, 2011, published in January, 2012. The spatial data for the vacant and derelict land was obtained through the Glasgow City Council's Development and Regeneration Services. The variables were all mapped and visually examined in relationship to the location of vacant and derelict land. [Figures 5 – 8]

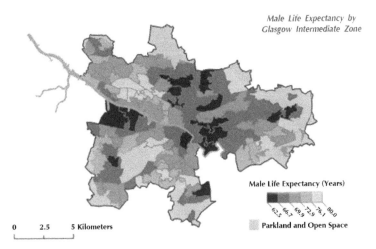

FIGURE 1.5   Male Life Expectancy (MLE) by Glasgow Intermediate Zone. Data Sources: U.K. Ordnance Survey (basemap layers); Scottish Neighbourhood Statistics, Scottish Government, 2010 (health data).

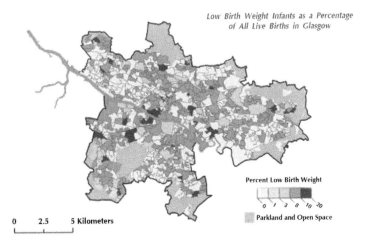

FIGURE 1.6   Low Birth-Weight Infants as a Percentage of All Live Births in Glasgow. Data Sources: U.K. Ordnance Survey (basemap layers); Scottish Neighbourhood Statistics, Scottish Government, 2010 (health data).

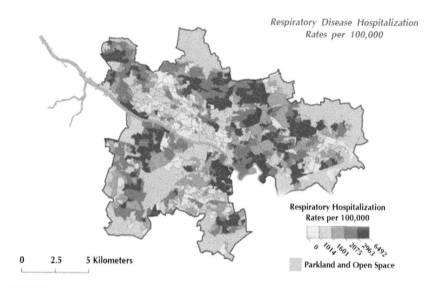

FIGURE 1.7   Respiratory Disease Hospitalization Rates per 100,000 by Glasgow Datazones. Data Sources: U.K. Ordnance Survey (basemap layers); Scottish Neighbourhood Statistics, Scottish Government, 2010 (health data).

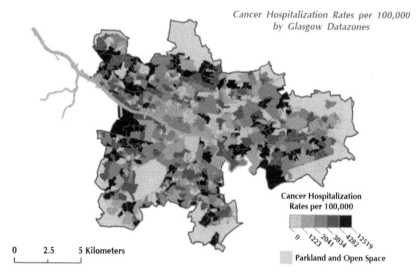

FIGURE 1.8   Cancer Hospitalization Rates per 100,000 by Glasgow Datazones. Data Sources: U.K. Ordnance Survey (basemap layers); Scottish Neighbourhood Statistics, Scottish Government, 2010 (health data).

1.2.3   Cluster Analysis, Geographically Weighted Regression, Descriptive Statistics, and Odds Ratios

After mapping the health variables and analyzing them visually, the cancer hospitalization rate dataset was selected as an example for further ESDA. Cluster analysis using Moran's I (Song and Kulldorff 2003), and Geographically Weighted Regression (Fotheringham 2002) were performed on the data in order to determine more specifically where inequities existed, where potentially anomalous high- and low-rate areas were located, and if any spatial patterns could be observed. [Figures 9 and 10]

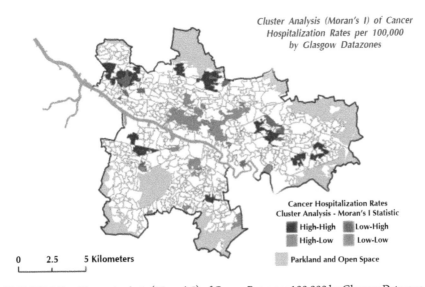

FIGURE 1.9   Cluster Analysis (Moran's I) of Cancer Rates per 100,000 by Glasgow Datazone. Data Sources: U.K. Ordnance Survey (basemap layers); Scottish Neighbourhood Statistics, Scottish Government, 2010 (health data).

In Figure 9, mapping the Moran's I clusters reveals areas which have high rates of cancer hospitalization surrounded by other high rate areas, and conversely, areas with low rates surrounded by other areas of low rates - in other words, the conditions which might be expected based on the principle of spatial autocorrelation, where "Everything is related to everything else, but near things are more related to each other than distant things," (Tobler 1970: 236). These, then, would be the concentrated clusters of high or low rates, (high-high or low-low, respectively). The anomalous areas are the areas of high rates surrounded by areas of low rates (high-low), and areas of low rates surrounded by areas of high rates (low-high). These are the areas which would potentially be fruitful to investigate further to find out why they are different from their immediate neighbors.

Similarly, the map in Figure 10 portrays the residuals of the Geographically-Weighted Regression (GWR), indicating the areas (in dark and medium orange) where the observed (actual) rates are one or more standard deviations above the rates that would be expected based on the relationship between cancer rates and the SIMD of that area. Conversely, the areas in dark and medium blue are one or more standard deviations below the overall rates of cancer hospitalization for Glasgow, indicating that these areas have a lower actual rate than would be expected based on the relationship between cancer rates and the SIMD.

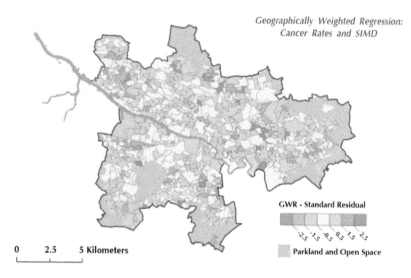

FIGURE 1.10   Geographically Weighted Regression (GWR) of Cancer Rates and SIMD. Data Sources: U.K. Ordnance Survey (basemap layers); Scottish Neighbourhood Statistics, Scottish Government, 2010 (health data); Scottish Index of Multiple Deprivation, General Report and Technical Report, Scottish Government Census, 2009 (SIMD data).

Subsequently, the 694 Data Zones of Glasgow were segmented into three classes, based on SIMD rank: High Deprivation DZs; Medium Deprivation DZs; and Low Deprivation DZs. [Figure 11] The descriptive statistics for each deprivation class were then calculated for each of the variables. Odds Ratios

(OR) were also calculated for the three health indicators for which data was consistent with the process of developing ORs.

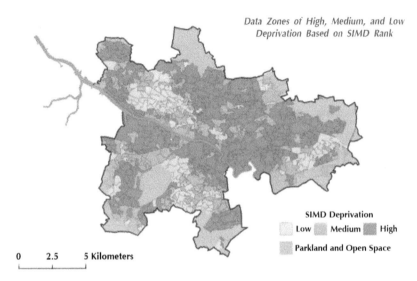

FIGURE 1.11   Datazones of High, Medium, and Low Deprivation, Based on SIMD Rank. Data Sources: U.K. Ordnance Survey (basemap layers); Scottish Index of Multiple Deprivation, General Report and Technical Report, Scottish Government Census, 2009 (SIMD data).

1.2.4   The Use of Vulnerability Indices in Environmental Hazard Assessment

In order to select Priority Areas for further analysis, and ultimately to recommend priority for these areas most in need of resource allocation and planning initiatives, an index, the Priority Areas for Re-Use of Derelict Land Index (PARDLI), was created to rank each DZ on the variables as analyzed, and then combine the ranks to obtain an overall score. The PARDLI scores thus reflect a combination of three categories: a measure of need, based on deprivation and social vulnerability through the inclusion of the SIMD; health vulnerability through the health outcomes indicators; and an exposure metric of proximity to VDL.

Traditionally, hazard assessment focused more on the physical exposure to the environmental hazard, with the assumption that any population in a hazardous area or exposed to a certain hazard, for example, would be equally impacted.

However, this approach does not include the very real additional risk of social vulnerability, meaning that populations that are disadvantaged through poverty, poor housing conditions, poor health, disability/special needs, lack of technology, age (either very young or elderly), racial/ethnic minority or immigration status, language barriers, and unemployment, would be more likely to be adversely affected by the same magnitude hazard as more affluent populations without those disadvantages.

"People's vulnerability is generated by social, economic, and political processes that influence how hazards affect people in varying ways and differing intensities….By 'vulnerability' we mean the characteristics of a person or group in terms of their capacity to anticipate, cope with, resist, and recover from the impact of a natural hazard," (Blaikie et al. 1994: 5, 9). Equity issues are of particular importance in risk assessment of environmental burdens and other natural and technological disasters. The socially and economically vulnerable, particularly if they have limited or no social support structure, may bear additional burdens than "mainstream" or more affluent populations when exposed to identical physical phenomena (Maantay and Maroko, 2009).

In addition to an area's population characteristics, differences in communities themselves also play a role. Place-based vulnerability, such as the degree of urbanization, population density, infrastructure, economic weakness, etc., can add to the inequities between places that translate into increased vulnerability in the face of the identical hazard (Mitchell 1999).

A number of indices have been developed in the past 20 years or so to help assess a population's vulnerability to hazards, most of which have focused on natural hazards such as flooding, hurricanes, and so forth. Vulnerability to environmental hazards is identified by assessing the potential exposure to the physical hazard; a measure of social resilience to the hazard; and an integration of the potential exposures and social resilience in a particular region.

One of the best-known vulnerability indices is the Social Vulnerability Index or SoVI (Cutter, Boruff, Shirley, 2003). This utilizes a hazards-of-place model of vulnerability. The SoVI was prepared on a county level for the entire United States, and utilizes census variables in a very similar way to the indicators included in the SIMD, as used in the PARDLI. After testing for multicollinearity amongst over 250 census variables, 42 variables were selected for principal components analysis, to reduce the index to 11 composite factors. The SoVI, as in the PARDLI, utilizes an additive model, so as not to make any subjective assumptions about the relative importance of the factors. Unlike the PARDLI, however, the SoVI is strictly a social vulnerability index and does not include any measures of exposure. Additionally, PARDLI is constructed at a finer resolution than the SoVI, allowing a more nuanced picture of vulnerability because the unit of analysis is much smaller, as is appropriate for a city-wide as opposed to a national level analysis.

Similar to the SoVI, the U.S. Centers for Disease Control and Prevention (CDC) has developed a Human Vulnerability Assessment model (HVA), which uses 15 U.S. census variables at the census tract level (US CDC 2008). The model calculates the percentile rank of ninety or higher for each variable. If a variable is within the ninetieth percentile, it receives a score of 1. The overall vulnerability is determined by summing the scores for all variables, which are not standardized and are weighted equally. The variables are similar to those used in the SoVI and the PARDLI, although the PARDLI represents an improvement as it includes an exposure component as well as more detailed health indicators.

Other researchers have adapted the HVA by modifying and augmenting the national-level index in order to construct a locationally-relevant vulnerability index, by incorporating geographically-specific datasets and including a biophysical component, which neither the HVA or the SoVI includes (Maantay, Maroko, and Culp 2010). In this way, the New York City Hazards Vulnerability Index (NYCHVI), which was designed for a specific location, taking into account the particularities of that location, and incorporating an exposure component, is most similar to the PARDLI.

Since the SoVI was first created, researchers have performed assessments and sensitivity analyses on vulnerability indices (Jones and Andrey 2007; Schmidtlein et al. 2008; Tate 2012). Although, in general, some differences were discerned when re-doing the analyses by altering indicator sets, analysis scale, methods of weighting, etc., it is likely that a one-size-fits-all approach (where the index performs equally well in every situation in every region) will be, by necessity, less than optimal. The best indices will be created for specific locations or regions, and reliant on expert guidance of those knowledgeable about the region.

"The quick, broad assessments of vulnerability provided by quantitative indices are useful guides for the selection of study areas in which more intensive, qualitative analyses may be conducted….Once a region's most vulnerable subareas are delineated in a systematic fashion, case-study research on the local drivers producing the pattern of vulnerability can begin, leading to reduced social vulnerabilities, and improved local resilience to environmental threats where and whenever they occur," (Schmidtlein et al. 2008: 1112). This type of "screening" index, to be used primarily for the selection of case-study areas for more detailed analyses, was the concept behind the creation of the PARDLI.

"Index development involves a multi-stage sequential process, which includes structural design, indicator selection, choice of analysis scale, data transformation, scaling, weighting, and aggregation. During each stage, modelers must make choices between multiple legitimate options…There is not necessarily a right, wrong, or best answer to these questions," (Tate 2012:327). With this in mind, the construction of the PARDLI is described below, acknowledging that, while there is a defensible rationale for all the decisions made, and the model was vetted with experts and their input was taken into account, no index will be a totally comprehensive reflection of reality, or a definitive predictor of risk, and there is always room for improvement and refinement. Undoubtedly, the PARDLI scores would have been calculated differently if different variables had been selected, if a different unit of analysis was used to aggregate the data, if different or additional environmental exposures were included, if the data had been classified in a different manner, or if the index had been weighted. The very nature of index creation entails that there will be assumptions made about which indicators best portray the conditions. The biggest limitations are that indicator data thought to be useful may not, in fact, exist or be available, and that not everything that should be included is known.

1.2.5 Development of PARDLI scores—Priority Areas for Re-use of Derelict Land Index

The PARDLI scores, intended to aid in the decision-making process of resource allocation, combine three aspects of vulnerability: overall high deprivation (need/social vulnerability), adverse health conditions (health vulnerability/ need), and proximity to an environmental burden (VDL) (exposure). The following section outlines how the Index was constructed.

Health variables (LBW, RESP, and CANCER) were re-classed into three categories: High, Medium, and Low, by classifying the rates and percentages by standard deviation. Numerical scores of 1, 2, and 3 were used to represent Low, Medium, and High, respectively. Any value above the first standard deviation over the mean received a "3" signifying the worst (or Highest rate) class. Any value below the first standard deviation below the mean received a "1" signifying the best (or Lowest rate) class. The middle group, between one standard deviation above and one standard deviation below the mean received a "2" signifying the middle (or Medium rate) class. For the LBW indicator, it was possible for a DZ to be assigned a score of "0," if the number of LBW infants for the years surveyed was zero. Nearly half of the DZs had no LBW infants, while in some DZs, as many as 1 in five infants were low-birth weight babies.

Male Life Expectancy data was given by IDZ, (Intermediate Data Zone, a larger unit than the DZ) so in order to be able to incorporate MLE into the Index, MLE values had to be assigned from the IDZ to the DZs within each larger zone. This is accomplished by spatially joining the DZs and the IDZs. The DZs nest hierarchically within the IDZs, so theoretically there should have been no overlap or gap issues. However, in the spatial database there were often slight boundary mismatches and therefore incorrect assignment of DZs to IDZs when spatially joining the polygons. In order to circumvent this problem, the centroids (points representing the geometric center of the polygon) of the DZs were spatially joined to the IDZ polygons instead. DZ centroids within a given IDZ were then assigned the MLE value of its parent IDZ, and the table containing the centroid values was joined to the original DZ polygon spatial database.

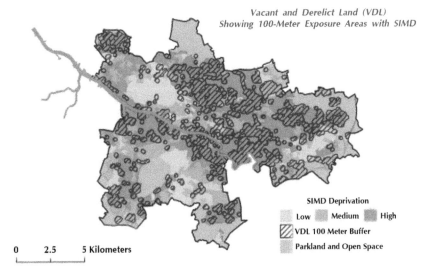

FIGURE 1.12   Vacant and Derelict Land, showing 100-meter Exposure Areas with SIMD. Data Sources: U.K. Ordnance Survey (basemap layers); Vacant and Derelict Land Survey, Scottish Government, 2012 (VDL data); Scottish Index of Multiple Deprivation, General Report and Technical Report, Scottish Government Census, 2009 (SIMD data).

Vacant and derelict land was buffered with 100-meter buffer distance, and any DZ that intersected one or more of these buffers was considered to be in proximity to a vacant and derelict land site. The 100-meter distance, rather than the 500-meter distance, was used, since so much of Glasgow was covered by the 500-meter buffers that the result would have been rendered almost meaningless. Additionally, 100 meters is a more conservative estimation of impact, from the standpoint of both visual blight and quality-of-life factors, as well as any potential impact from contamination. This metric appears in the Index as a binary feature: the DZ is either proximate, in which case it received a score of "3," or not proximate, in which case it received a "0," to a vacant or derelict site. [Figure 12]

In the absence of any compelling rationale for weighting one variable higher than the others, the Index was created by simple addition of the six variables' scores. Combined PARDLI scores ranged from a low of 4 (best, lowest priority area) to a high of 18 (worst, highest priority area). A score of "4" indicates that the DZ has the lowest (best) scores possible for all variables: four "1"s and two "0"s. A score of 18 indicates that the DZ has the highest (worst) possible scores for all six variables: six "3"s. The scores were then divided into the three classes of Low, Medium, and High, as before for the individual variables.

1.2.6 Selection of Case Study Priority Areas

The intention of developing the PARDLI scores is to select the areas with the highest need on an objective, quantitative basis. [Figure 13] These areas would presumably have the highest deprivation scores, worst health outcomes scores, and be proximate to vacant and derelict land. The following neighborhoods (using the Intermediate Data Zone boundaries as representing more accurately actual neighborhoods than do the individual DZs) meet those criteria: from west to east in Glasgow: Drumchapel South; Govan-Linthouse; Possil Park; CaltonGallowgate-Bridgeton; and Old Shettleston-North Parkhead. [Figure 14]

The selected Priority Areas will be the focus of a more detailed analysis, to consider the role of historic land use and settlement patterns; possibly more detailed health data from surveys rather than just aggregated statistics; a quality assessment of parks and open space; existing and proposed development initiatives; and existing community organizations and activities taking place in each neighborhood. Although there is a flurry of diverse planning initiatives ongoing or in the development stages for Glasgow, these appear to be somewhat disjointed, with perhaps some gaps, overlaps, and conflicts amongst them. They might benefit from a more unified focus and implementation strategy. [Figure 15]

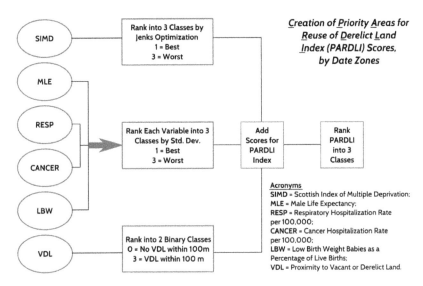

FIGURE 1.13 Flow Diagram Depicting the Development of the PARDLI Index

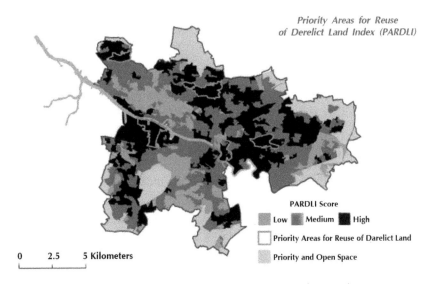

FIGURE 1.14 Priority Areas for Reuse of Derelict Land Index (PARDLI). Areas outlined within red boundaries, from west to east in Glasgow: Drumchapel South; Govan-Linthouse; Possil Park; Calton-GallowgateBridgeton; and Old Shettleston-North Parkhead. Data Sources: U.K. Ordnance Survey (basemap layers); Maantay (PARDLI scores, see Figure 13 for datasets included)

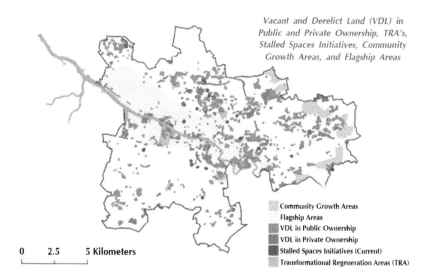

Vacant and Derelict Land (VDL) in Public and Private Ownership, TRA's, Stalled Spaces Initiatives, Community Growth Areas, and Flagship Areas

Community Growth Areas
Flagship Areas
VDL in Public Ownership
VDL in Private Ownership
Stalled Spaces Initiatives (Current)
Transformational Regeneration Areas (TRA)

0   2.5   5 Kilometers

FIGURE 1.15   Vacant and Derelict Land in public and private ownership, Transformational Regeneration Areas (TRAs), Stalled Spaces Initiatives, Community Growth Areas, and Flagship Areas. Data Sources: U.K. Ordnance Survey (basemap layers); Vacant and Derelict Land Survey, Scottish Government, 2012 (VDL data); Glasgow City Council Development and Regeneration Services; Glasgow and Clyde Valley Green Network Partnership.

For instance, various agencies and committees have designated areas within Glasgow and the Greater Clyde Valley region as "Transformational Regeneration Areas (TRAs), "Community Growth Areas," and "Metropolitan Flagship Areas." In addition, there are the property lots associated with the "Stalled Spaces" Initiative, which promotes the temporary use of vacant lots by the community that may have previously been slated for development of one type or another, but now are lying fallow until the developer (usually a private owner) decides that the time is right to proceed with the original plans (Glasgow City Council 2013a). The objective of the Transformational Regeneration Areas, according to the Glasgow City Council's website, is to transform disadvantaged neighborhoods by major re-structuring and to retain the current community and to attract people back into these areas (Glasgow City Council 2013b). The program is not just for housing but is intended to also deliver local opportunities such as jobs, education, training, and community facilities, and is considered one of the most ambitious programs of urban renewal in the U.K. (Glasgow and Clyde Valley Structure Plan Joint Committee 2006; Fletcher 2011). Community Growth Areas have been identified by the Glasgow and Clyde Valley Structure

Plan Joint Committee as primarily residential areas in need of expanded master urban planning efforts, especially additional housing opportunities, along key transportation corridors (Glasgow and Clyde Valley Structure Plan Joint Committee 2006). Metropolitan Flagship Areas "are central to the restructuring of the Metropolitan Area and to the competitiveness of Scotland. They also continue to offer opportunities to accommodate major investment, for example associated with the bid for the Commonwealth Games 2014. The National Planning Framework and Regeneration Policy Statement recognises the Clyde Waterfront, Clyde Gateway and Ravenscraig as national priorities for regeneration and renewal," (The Scottish Government Directorate for Planning and Environmental Appeals 2012). The Flagship Initiatives, in particular, support the development of green networks and reuse of brownfields

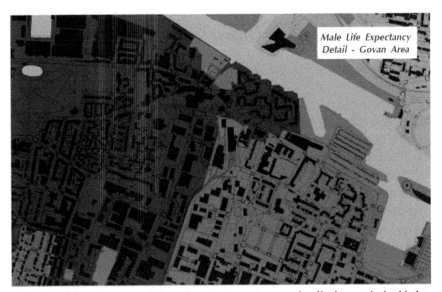

FIGURE 1.16 The proposed Govan Priority Area—An example of looking at the health data at the local level. Brown areas are the VDL. The darker the blue, the lower the male life expectancy (MLE). Most of Govan is in the worst or second worst MLE class, out of the original five classes mapped. Data Sources: U.K. Ordnance Survey (basemap layers); Vacant and Derelict Land Survey, Scottish Government, 2012 (VDL data); Scottish Neighbourhood Statistics, Scottish Government, 2010 (health data).

A detailed perspective of the specific Priority Areas, drilling down in the data to the largest possible scale, or "looking through the spatial microscope," is an important step in developing effective policy recommendations and implementation schemes. Basic science enables the generation of general principles, which are likely place-less. But the application of scientific knowledge to policy, especially to local policy, requires a thorough understanding of spatial variation at the local level, at a high level of resolution, and is rooted in "placeness;" thus, the necessity of delving deeper into the micro-environment to ferret out the impact of neighborhood effects (Goodchild et al. 2000). [Figure 16]

"Place is not merely a setting or backdrop, but an agentic player in the game – a force with detectable and independent effects on life," (Werlen 1993, as paraphrased in Gieryn 2000:466). Researchers typically use such units of analysis as census tracts or data zones, populated with socio-demographic data. But these, as such, are not places, merely a convenient area in which to bundle variables. In order to be "place-sensitive," we need to include information about "the relative location of the census tract [or datazone] within the metropolitan area, the pattern of streets or significance of particular buildings, such as churches or markets, and the perceptions and understandings of the place by the people who might live there or not," (Gieryn 2000:466).

This more detailed exploration of "place," "place-ness," and "emplacement" is the purpose of selecting the case-study areas for further analysis. The next phase of this research is to conduct a three-pronged approach: 1) an analysis of historical conditions in the PARDLI neighborhoods, particularly regarding the prior land uses of the VDL; 2) more detailed health analysis; and 3) incorporation of the quality assessment of the parks and open spaces. Additionally, we will be surveying the communities as to extent of social capital as evidenced by community-led organizations, neighborhood improvement activities, influence on local politics, and community cohesiveness. The potential for PARDLI neighborhoods to respond to problems of deprivation is an important aspect of the future research. An investigation into the existing community organizations and activities in each PARDLI area will help to evaluate how these organizations might be able to mitigate the negative impacts of the VDL.

Although individual behavioral factors undoubtedly account for some of the poor health of Glasgow's most deprived populations, neighborhood effects are also an important consideration, and include such influences as differential exposure to stressors and differences in social infrastructure (Diez-Roux 2001; MacIntyre 2002; Kawachi and Berkman 2003; Croucher et al. 2007). Scotland's Chief Medical Officer, Sir Harry Burns, believes that these stressors and their concomitant health impacts go a long way in explaining the poor health here (Burns 2012).

1.2.7 *Greenspace Analyses*

Many research studies have examined the relationship between access to open space and health benefits, and although the link has not been definitively and consistently demonstrated, a number of studies have found correlations between health benefit and access to open space and areas promoting physical activity (De Vries et al. 2003; Bedimo-Rung et al. 2005; Giles-Corti et al. 2005; GordonLarsen et al. 2006; Groenewegen et al. 2006; Maas et al. 2006; Roemmich et al. 2006; Diez-Roux et al. 2007; Mitchell and Popham 2007; Mitchell and Popham 2008; Rundle et al. 2008). Ideally, some measure of access to or amount of greenspace for each DZ could have been included in the PARDLI Index. After all, if we are thinking about creating priority areas for re-use of vacant and derelict land for the possible augmentation of existing greenspace, or in some way to compensate for the lack of accessible greenspace, this would have been a logical indicator to have incorporated. However, there are some rather unique factors involved with the quantity and distribution of Glasgow's existing greenspace, as well as some more typical problems of arriving at a true estimate of greenspace access, which are the same in an analysis of any city's greenspace.

Glasgow is extremely well-endowed with parks and other publiclyavailable open space. There is a sizable greenbelt area which nearly encompasses the perimeter of the city, and several large parks are centrally located throughout the city. Additionally, there are myriad other categories of designated open space and active recreational facilities, as well as a significant quantity of land protected as natural habitat areas. There are 33 separate categories of public open space designated, including parks, gardens, sports areas, amenity spaces within developments, green corridors, protected natural areas, nature reserves, historic landscapes, and ancient woodlands. Indeed, when we look at a map with all these classes of open space plotted out, Glasgow is practically covered with greenspace of one kind or another. [Figure 17]

Several comprehensive analyses by other researchers have explored access to greenspace and greenspace quantities in Glasgow. The Center for Research on Environment, Society, and Health (CRESH) at the Universities of Edinburgh and Glasgow developed a model to predict percentage of open space in each ward of the entire United Kingdom (Pearce et al. 2008; Richardson et al. 2010; Shortt et al. 2011). [Figure 18]

When extracting and mapping just the Glasgow wards from their data, one can see that the only wards to have less than 20% of their area in open space are

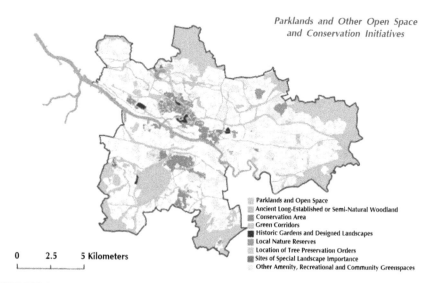

FIGURE 1.17 Glasgow's Open Space. There are 33 separate categories of public open space designated. Data Sources: U.K. Ordnance Survey (basemap layers); Planning Advice Notice (PAN) 65 Planning and Open Space, Scottish Government, 2008, (open space data).

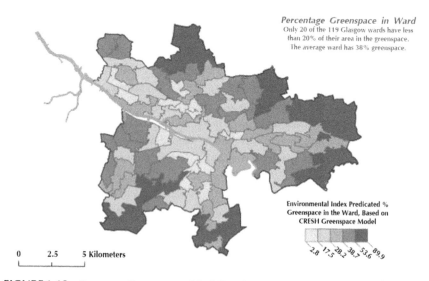

FIGURE 1.18 Percentage Greenspace in Ward. Data Sources: U.K. Ordnance Survey (basemap layers); Developing Summary Measures of Health-Related Multiple Physical Environmental Deprivation for Epidemiological Research, Richardson, et al., 2010 (CRESH model data)

the more highly urbanized parts of the city centre, and indeed the wards have on average 38% of their area in open space. A few wards have nearly 90% of their areas in open space, which is an extraordinarily high figure, and these wards tend to be near the peripheral areas and in some of the more deprived areas of the city.

A separate analysis, "Networks for People," conducted by the Glasgow and Clyde Valley Green Network Partnership, was intended to show "connectedness" to greenspace by using actual network walking distance from each property lot to the greenspace entrance, taking into account physical barriers such as motorways and rivers. The city was divided up into a tessellation of 100- meter hexagonal cells, and a value assigned to each cell, indicating the degree of connectivity, based on the network analysis. [Figure 19] The white cells on the map indicate excellent connectivity, and the darker the color, the worse the connectivity. The vast majority of the cells show very good connectivity, with some patches of disconnectedness, again, as with the CRESH analysis, most prevalent in the more densely built-up centre city areas, which in many cases correspond to the more affluent parts of the city (Glasgow and Clyde Valley Strategic Development Planning Authority 2011).

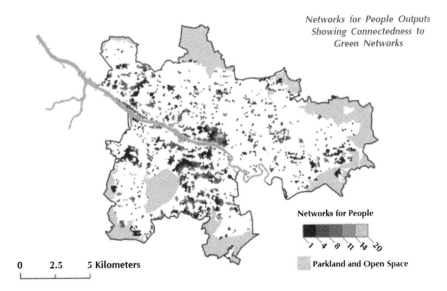

FIGURE 1.19   Networks for People Outputs, showing connectivity to Greenspace. The lower the NfP score, the more disconnected that 100 m cell is from the Green Network (model output of green network). Data Sources: U.K. Ordnance Survey (basemap layers); Glasgow and Clyde Valley Green Network Partnership, 2011.

Based on these two analyses and my own observations of the greenspace data from the Planning Advice Notice 65 (PAN 65) and Integrated Habitat Network (IHN) datasets (Scottish Government 2008; Smith et al. 2008), Glasgow appears well-provisioned with greenspace, and moreover, the less affluent areas often have a higher proportion of greenspace and in closer proximity than the more affluent areas. However, neither of these analyses takes into account the quality or usability of the greenspace for any beneficial purpose (Maroko et al. 2009; Miyake et al. 2010). Oftentimes the so-called greenspace is little more than a dumping ground for old sofas and rubbish, or else is viewed as a dangerous place to the local residents, who do not make use of it. In some cases it is just an impassible overgrown area with no amenities, or for some other reason is not user-friendly. There is a survey currently being undertaken to assess greenspace quality, but it is only partially completed at this time. Without an assessment of quality, it would be very difficult to base any greenspace access score on geographic access or quantity of greenspace alone, and therefore this would be rather meaningless as an indicator to incorporate into the PARDLI scores. However, since this inventory is a work in progress, it is hoped that by the time the Priority Areas are looked at in a detailed format, the greenspace quality data will be available for use.

1.3   FINDINGS

The spatial analysis of disease and other health metrics by Data Zones within Glasgow shows that some neighborhoods, and therefore some populations, suffer from poor health and low life expectancy disproportionately more than others. Many of these areas correspond spatially to areas of high deprivation and areas with excessive vacant and derelict land—often former environmentally-noxious land uses—making these populations vulnerable in more ways than one. The areas of highest deprivation and worst health deciles spatially correspond almost totally with the location of the VDL.

When looking at the results of the cluster analysis using Moran's I for cancer hospitalization rates as an example, the clusters of DZs having high rates that are surrounded by other DZs with high rates also correspond to these areas of particularly high deprivation. Likewise with the GWR analysis, in some areas the regression models predict much lower rates than the actual observed values. In certain DZs the observed rate of cases is more than 2.5 standard deviations above the predicted, based on the regression relationship with deprivation rank, and therefore the reality of the cancer hospitalization rates in these areas is much worse than what would be predicted based on deprivation alone.

TABLE 1.1   Descriptive Statistics for High, Medium, and Low Deprivation Areas in Glasgow

SIMD Data Zones	# Data Zones	Population	Vacant & Derelict Hectares per 1,000 Pop	% Total VDL Hectares	Cancer Hospitalization Rates/ 100,000	Respiratory Hospitalization Rates/ 100,000	% Low Birth Weight of Total Live Births	MLE by IDZ	Male Life Expectancy Range
High Deprivation	375	298,224	3.7	69	3,807	2,571	3.55	66.5	62.5-74.7
Medium Deprivation	180	139,325	2.5	23	2,852	1,637	2.87	73.2	67.4-77.4
Low Deprivation	139	99,775	1.2	8	2,781	999	1.55	75.8	69.9-80.0
SUM/AVG.	694	537,324	2.4 Mean	100%	2,872 Mean	2,014 Mean	2.9 Mean	71.3 Mean	62.3-80.0

The descriptive statistics show the difference in amount of vacant and derelict land, health outcomes, and life expectancy, as differentiated by High, Medium, and Low Deprivation DZs. [Table 1; Figures 20-23] As shown in the tables and graphs, there are 3.7 hectares of vacant and derelict land per 1,000 people in High Deprivation DZs, as opposed to 1.2 hectares of VDL per 1,000 people in Low Deprivation DZs. Likewise, Male Life Expectancy averages 66.5 years in High Deprivation DZs, while it is 75.8 years on average in Low Deprivation DZs. Hospitalization rates for cancer and respiratory disease and the percentage of low birth weight infants are correspondingly much higher in High Deprivation DZs, as well.

Percent Vacant and Derelict Land

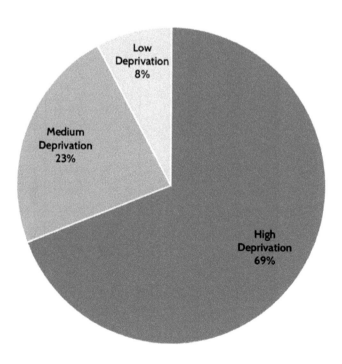

FIGURE 1.20   The Breakdown of Vacant and Derelict Land in Glasgow, by High, Medium, and Low Deprivation Areas

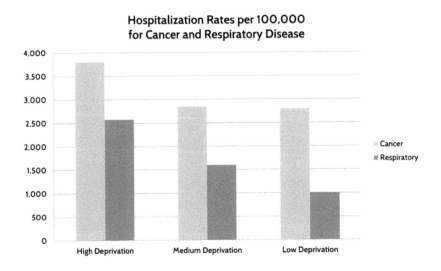

FIGURE 1.21   Hospitalization Rates per 100,000 in Glasgow for Cancer and Respiratory Diseases, by High, Medium, and Low Deprivation Areas

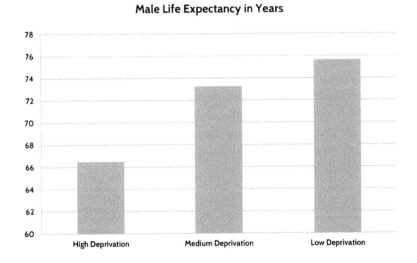

FIGURE 1.22   Male Life Expectancy in Glasgow by High, Medium, and Low Deprivation Areas

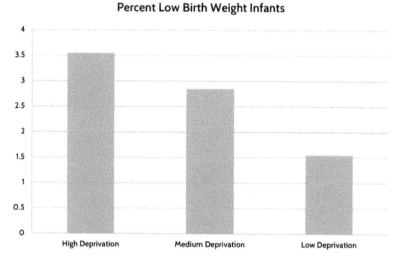

FIGURE 1.23   Percentage of Low Birth Weight Infants in Glasgow, by High, Medium, and Low Deprivation Areas

By calculating the Odds Ratios, it can be seen that the differences in these health variables between the High Deprivation DZs and the other DZs are statistically significant. [Table 2] Calculating Odds Ratios is a way of comparing data from two different populations in order to obtain a quantitative evaluation of real significance in the differences between the two groups. Odds Ratios are a surprisingly simple, yet powerful way to show statistical associations in health. They are particularly helpful in demonstrating health inequalities. The Odds Ratio is the odds of disease or health outcome among exposed individuals (in this case, people living in a High Deprivation DZ) divided by the odds of the disease or health outcome among the unexposed (in this case, people living in a DZ that is not High Deprivation).

TABLE 1.2   Odds Ratios for Health Outcomes in Glasgow in High Deprivation Areas 95% CI, with p = < 0.0001.

Health Outcome	Odds Ratio
Respiratory Hospitalization	5.5
Cancer Hospitalization	1.3
Low Birth Weight Infants	1.6

Results of the OR analysis show that populations in High Deprivation DZs are much more likely to be hospitalized for respiratory disease (5.5 times more likely) or cancer (30% more likely), and much more likely to have low birth weight infants (60% more likely), than those not living in High Deprivation DZs. The analysis of risk factors for unfavorable health outcomes is based on a comparison between cases and non-cases in High Deprivation DZs, and cases and non-cases in non-High Deprivation DZs. All results are at the 95% Confidence Level, with $p = < 0.0001$.

The creation of the PARDLI results in five areas in the highest scoring categories for deprivation, health outcomes, and proximity to VDL. There is a high degree of spatial correspondence between the areas with concentrations of VDL and the DZs with the highest PARDLI scores. [Figure 24]

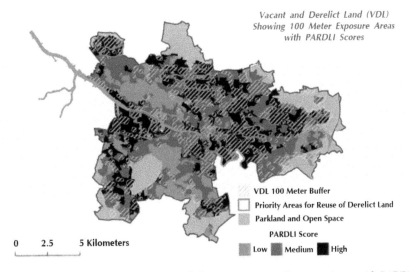

FIGURE 1.24   Vacant and Derelict Land showing 100-meter Exposure Areas with PARDLI Scores. Data Sources: U.K. Ordnance Survey (basemap layers); Vacant and Derelict Land Survey, Scottish Government, 2012 (VDL data); Scottish Index of Multiple Deprivation, General Report and Technical Report, Scottish Government Census, 2009 (SIMD data).

1.4   CONCLUSIONS, RECOMMENDATIONS, AND FUTURE DIRECTIONS

The PARDLI scores appear to accurately reflect areas within Glasgow that 1) are exposed to the blight and potential environmental burdens of vacant and derelict land; 2) have high levels of social vulnerability; and 3) have high needs due to poor health conditions. Whilst the PARDLI approach can be transferred to other locations, we believe that vulnerability indices such as PARDLI are perhaps better used as guidelines for other locations, and should be adjusted, if used elsewhere, to better accommodate local conditions and issues. As mentioned earlier in this paper, broad-brush indices are best if utilized mainly as a "screening" device for selecting case-study areas for more detailed study. This paper has discussed the development, assumptions, and limitations of such an approach, as an illustration of index creation for a specific location.

The five Priority Areas as determined by the highest PARDLI scores will be analyzed further in terms of incorporating more detailed data in order to make cogent community-specific recommendations regarding the reuse of the VDL within each of these areas, as mentioned above. The methodology described in this paper can be applied to any other local authority in Scotland, and indeed, to any location where the VDL, health, and deprivation data exist. In countries not using a standardized index of deprivation, other salient variables could be substituted as proxies for deprivation, such as income or poverty levels.

For policy and planning initiatives, we need to start thinking differently about how best to serve these communities that have ended up in this analysis with the highest PARDLI scores. Perhaps it is worthwhile to examine the differences between re-use of VDL for "regeneration," versus using it for "development," and think more seriously about who usually benefits most from regeneration or development. In the parts of the city where VDL is most prevalent, it seems unlikely that there will be high interest from private investors to construct profit-making facilities (i.e., "development"). It might be better to acknowledge this and move on to realistic re-use concepts for the VDL, and plan for uses that would more directly benefit the surrounding community and serve their needs directly, as opposed to being held for a regional use or general taxgenerating purpose. Regeneration for primarily community use can result in substantial gains in many aspects, including health and other more nonquantifiable benefits. It can also have an economic multiplier effect and may serve to bolster the local economy, and even have ripple and spill-over effects to neighboring communities. Regeneration of this nature should not be discounted just because it does not involve constructing a commercial or residential building complex. There are also valid reasons for not encouraging housing to be built on VDL sites, as argued by Greenberg in "Should housing be built on former brownfield sites?" (Greenberg 2002).

It is important for the community to not only participate and be involved in the decision-making process, but to actually take the lead on devising plans and implementation strategies for the VDL. This needs to be a bottom-up planning initiative, not one led by professional planners, but rather one optimizing community leadership while drawing on planning expertise. This will better ensure community satisfaction with the eventual project, as well as serve to bolster capacity building in the community, so that the end product can be selfsustaining and successful, and engender a sense of community ownership and a source of local pride.

More than half of the VDL sites in Glasgow are in public ownership (572 out of 927 sites), representing approximately 783 out of the 1,300 total hectares of VDL. This means that Glasgow city government could effectively grant highly deprived communities more than 700 hectares of land to be used for community good. This might be urban agriculture in the form of communal gardens (as opposed to individual "allotments"), provided that urban agriculture is appropriate for that community, as determined by the community itself, and of course using raised bed planting methods in areas likely to have contaminated soil. In New York City, and in other cities around the world, community gardens have proved to be an effective way to get some very positive constructive results without substantial financial outlays, and the health and other benefits of community gardens and urban agriculture have been well-documented (Armstrong 2000; Schukoske 2000; Holland 2004; Ottmann et al. 2010; MacKeen 2011; Ottmann et al. 2012).

Benefits include:

- Improvements in community cohesiveness and neighborliness;
- Increases in healthy food options, especially where highly deprived communities are likely to be "food deserts";
- Expansion of environmental awareness for children and youths;
- Provision of a strong geographic focal point for community cultural and educational activities;
- Improvements in neighborhood aesthetics;
- Enhancement of property values;
- Reduction in crime rates, due to more "eyes on the street," increased pride and involvement in the neighborhood by residents, and created constructive opportunities and activities for children and youth;
- Development of community participation in other important issues, and energizing their activities.
- Promotion of community building, capacity building.

Policy and planning recommendations for re-use of VDL by community members or organizations can include such strategies as:

- Creating a database of publicly-owned vacant sites that are accessible from the street, and making this list available to the public;
- Developing a signage program for each of these sites advising community members who to call to discuss community-led use of the site. These could be simple hand-lettered signs, similar to the ones put up on VDL sites in NYC by the non-profit group known as 596 Acres (which refers to the amount of VDL in Brooklyn, NYC) (Leland 2012; 596 Acres 2012).
- Establishing a standard protocol for leasing the land to a community group, and have a small support team within government to help with logistics of community-led use of the vacant land;
- Thinking more flexibly about appropriate uses, whether temporary or permanent. Community uses could be urban agriculture, passive or active recreation spaces, market spaces for weekly "flea" markets or farmers' markets, and cleaned up natural areas that might connect with other open space networks;
- Allow and facilitate true community planning. Rather than top-down planning for the vacant space, community participation (and even community initiation of the project) at the earliest stages would be more likely to ensure community "buy-in" to the decisions and community involvement in the continued success of the use to which the land is put;
- Consider small grants of money for community-led groups to create containerized gardening on sites that may be contaminated, and that can be moved to another vacant site if the gardening site is eventually required for brownfields remediation and housing development;
- Using the land temporarily for urban forestation projects. These urban forestry plantings could help clean up contamination through phytoremediation, help restore endangered tree species, and create economic benefit, while leaving land available for future housing development or other community use. Urban forestation can also be a permanent use (Eadha Enterprises 2012).

This research is part of a larger project involving a comparison of Glasgow and New York City (NYC) regarding the relationship between environmental health justice and the built environment. Preliminary findings of the NYC analysis reveal that similar conditions exist regarding the approximate percentage of city land in the vacant and derelict category (4% for Glasgow, and 5%

for NYC) and also that the land is disproportionately located in the lessaffluent communities in both cities. It is expected that at the end of the comparison study some conclusions can be drawn about best management practices and strategies for use of the VDL, drawing upon what has been found to work in these cities. The environmentally-responsible re-use of vacant land is an important issue for many post-industrial cities, and therefore the findings and recommendations have wide application beyond the two case study cities, especially in those urban areas with severe inequities in economic power, health outcomes, opportunities for healthy active living, and other quality of life concerns.

As in Glasgow, much of New York City's vacant land is located in the poorer neighborhoods. A major issue in NYC with re-use of vacant and derelict land for development is the displacement of poor people through gentrification. Ironically, this has often occurred in areas where community gardens have improved property values, enhanced neighborhood aesthetics, and reduced crime rates sufficiently to interest developers in investing in the neighborhood, whereby the community rightfully feels as though their hard work has sown the seeds of their own destruction (Smith and DeFilippis1999; Von Hassel 2002). Policies must be in place for community-led improvements in vacant and derelict land to benefit the community and not punish them.

Actively promoting the re-use of vacant and derelict land in high deprivation areas with vulnerable populations will have long-term beneficial use to the residents, and is an important step in combating health inequities and environmental injustice in these communities.

"[A] society that allows such a pattern of coincidence [between poor populations and poor environment] to persist has failed to equally protect its citizens. This failure, itself, constitutes an environmental injustice. Whether the result of…putting economic profits over the health of people, or benign neglect, this disproportionate risk can and does lead to disastrous results," (White 1998:75).

REFERENCES

1. Armstrong, D. (2000). A survey of Community Gardens in Upstate New York: Implications for Health Promotion and Community Development. Health and Place, 6(4), 319-317.
2. Baibergenova, A., Kudyakov, R., Zdeb, M., Carpenter, D.O. (2003). Low birth weight and residential proximity to PCB-contaminated waste sites. Environmental Health Perspectives, 111(10):1352-7.
3. Bedimo-Rung, A.L., A.J. Mowen, and Cohen, D.A. (2005). The significance of parks to physical activity and public health: A conceptual model. American Journal of Preventative Medicine, 28(2S2):159-168.

4. Blaikie, P., Cannon, T., Davis, I., and Wisner, B. (1994). At Risk: Natural Hazards, People's Vulnerability, and Disasters. London: Routledge.

5. Brack, C. L. (2002). Pollution mitigation and carbon sequestration by an urban forest. Environmental Pollution, 116(S1):195–200.

6. Brender, J., Maantay, J., and Chakraborty, J. (2011). Residential Proximity to Environmental Hazards and Adverse Health Outcomes. American Journal of Public Health, 101(S1): S37-S52.

7. Bryant, B., ed. (1995). Environmental Justice: Issues, Policies, and Solutions. Washington, DC: Island Press.

8. Bullard, R.D., ed. (1994). Unequal Protection: Environmental Justice and Communities of Color. San Francisco: Sierra Club Books. Burns, H. (2012). Presentation at Heads of Planning Scotland (HoPS) Annual Meeting, The Lighthouse, Glasgow, June 15, 2012.

9. Chadwick, E. (1842, reprinted 1965). Report on The Sanitary Condition of the Labouring Population of Great Britain. Edinburgh: University of Edinburgh Press.

10. Chakraborty, J., and Armstrong, M.P. (1997). Exploring the use of buffer analysis for the identification of impacted areas in environmental equity assessment. Cartography and Geographic Information Systems, 24(3):145–157.

11. Craig, C. (2010). The Tears that Made the Clyde: Well-Being in Glasgow. Argyll: Argyll Publishing.

12. Crawford, F., Beck, S., and Hanlon, P. (2007). Will Glasgow Flourish? Regeneration and Health in Glasgow: Learning from the past, analyzing the present, and planning for the future. Glasgow: Glasgow Centre for Population Health.

13. Croucher, K., Myers, L., Jones, R., Ellaway, A., and Beck, S. (2007). Health and the Physical Characteristics of Urban Neighbourhoods: A Critical Review. Glasgow: Glasgow Centre for Population Health.

14. Cutter, S.L., Boruff, J., Shirley, W.L. (2003). Social Vulnerability to Environmental Hazards. Social Science Quarterly, 84(2): 24-261.

15. De Vries, S., Verheij, R. A., Groenewegen, P. P., & Spreeuwenberg, P. (2003). Natural environments – Healthy environments? An exploratory analysis of the relationship between greenspace and health. Environment and Planning A, 35(10), 1717–1731.

16. Diez-Roux, A.V. (2001). Investigating Neighborhood and Area Effects on Health. American Journal of Public Health, 91(11):1783-1789.

17. Fotheringham, A.S., Brunsdon, C., Charlton, M. (2002). Geographically weighted Regression: the analysis of spatially varying relationships. West Sussex, England: John Wiley & Sons Ltd.

18. Fraser, W. H. and Maver, I., (eds). (1996). Glasgow 1813 to 1912. Manchester: Manchester University Press.

19. Fullilove, M.T., Fullilove, R.M. III. (2000). Place matters. In: Reclaiming the Environmental Debate: The Politics of Health in a Toxic Culture. Hofrichter, R., ed. Cambridge, MA: Massachusetts Institute of Technology, 77–91.

20. Gibb, A. (1983). Glasgow. The Making of a City. London: Routledge.

21. Gieryn, T.F. (2000). A space for place in sociology. Annual Review Sociology, 26(1): 463-497.

22. Giles-Corti, B., Broomhall, M., Knuiman, M., Collins, C., Douglas, K., Ng, K., Lange, A., Donovan, R. (2005). Increasing walking: How important is distance to, attractiveness, and size of public open space? American Journal of Preventive Medicine, 28(2S2):169-176.

23. Glasgow and Clyde Valley Strategic Development Planning Authority (2010). Vacant and Derelict Land Monitoring Report 2010. Glasgow: Glasgow and Clyde Valley Strategic Development Planning Authority.

24. Glasgow and Clyde Valley Strategic Development Planning Authority (2011). Green Network Spatial Priorities. Glasgow: Glasgow and Clyde Valley Strategic Development Planning Authority.

25. Glasgow and Clyde Valley Structure Plan Joint Committee (2006). Glasgow and Clyde Valley Joint Structure Plan 2006 Supplementary Written Statement Third Alteration to the 2000 Plan at http://www.gcvcore.gov.uk/DOCS/structure_plan/2006_Plan_Supplementary_Written_Statement.pdf (last accessed July 23, 2013)

26. Glasgow Centre for Population Health (2008). A Community Health and Wellbeing Profile for East Glasgow. Glasgow: Glasgow Centre for Population Health.

27. Glasgow City Council (2013a). Glasgow Stalled Spaces at http://www.glasgow.gov.uk/stalledspaces and http://www.glasgow.gov.uk/index.aspx?articleid=5118 (last accessed July 23, 2013)

28. Glasgow City Council (2013b). Glasgow's Transformational Regeneration Areas (TRAs):Transforming Disadvantaged Neighbourhoods. at http://www.glasgow.gov.uk/index.aspx?articleid=7533 (last accessed July 23, 2013)

29. Goodchild, M., Anselin, L., Appelbaum, R.P., and Harthorn, B. (2000). Toward Spatially Integrated Social Science. International Regional Science Review, 23(2):139-159.

30. Gordon-Larsen, P., Nelson, M.C., Page, P., and Popkin, B.M. (2006). Inequality in the built environment underlies key health disparities in physical activity and obesity. Pediatrics, 117:417-424.

31. Gray, L. (2008). Comparisons of health-related behaviours and health measures in Greater Glasgow with other regional areas in Europe. Glasgow: Glasgow Centre for Population Health. Greenberg, M., Lee, C., Powers, C. (1998). Public Health and Brownfields: Reviving the Past to Protect the Future. American Journal of Public Health, 88 (12):1759-1760.

32. Greenberg, M. (2002). Should housing be built on former brownfield sites? American Journal of Public Health, 92: 703-5.

33. Grineski, S.E., Collins, T.W., Chakraborty, J., and McDonald, Y. (2013). Environmental Health Injustice: Exposure to Air Toxics and Children's Respiratory Hospital Admissions in El Paso, Texas. The Professional Geographer, 65 (1):31-46.

34. Groenewegen, P. P., van den Berg, A. E., De Vries, S., & Verheij, R. A. (2006). Vitamin G: Effects of green space on health, well-being, and social safety. BMC Public Health, 6:149–158.

35. Hanlon, P., Lawder, R.S., Buchanan, D., Redpath, A., Walsh, D., Wood, R., Bain, M., Brewster, D.H. and Chalmers, J. (2005). Why is mortality higher in Scotland than in England & Wales? Decreasing influence of socioeconomic deprivation between 1981 and 2001 supports the existence of a 'Scottish Effect'. Journal of Public Health, 27(2):199-204.

36. Hofrichter, R. (ed.) (1993). Toxic struggles: The theory and practice of environmental justice. Philadelphia, PA: New Society Publishers.

37. Holland, L. (2004). Diversity and Connection in Community Gardens: a contribution to local sustainability. Local Environment, 9(3): 285-305.

38. Horsey, M. (1990). Tenements and Towers: Glasgow Working-Class Housing, 1890-1990. Edinburgh: Royal Commission on the Ancient and Historical Monuments of Scotland.

39. Hume, J. R. and Moss, M. (1977). Workshop of the British Empire. London: Heinemann.

40. Johnston, B.R., ed. (1994). Who Pays the Price? The Sociocultural Context of Environmental Crisis. Washington, DC: Island Press.

41. Jones, B., and Andrey, J. (2007). Vulnerabilty index construction: methodological choices and their influence on identifying vulnerable neighborhoods. International Journal of Emergency Management, 4(2):269-295.

42. Kawachi, I., and Berkman, L.F., eds. (2003). Neighborhoods and Health. New York: Oxford University Press.

43. Kuehn, C.M., Mueller, B.A., Checkoway, H., Williams, M. (2007). Risk of malformations associated with residential proximity to hazardous waste sites in Washington State. Environmental Research, 103(3):405-412. Leland, J. (2012). Turning Unused Acres Green. The New York Times, April 27th, 2012.

44. Litt, J. and Burke, T. (2002). Uncovering the historic environmental hazards of urban brownfields. Journal of Urban Health, 79(4): 464-81.

45. Litt, J., Tran, N., and Burke, T. (2002). Examining Urban Brownfields through the Public Health "Macroscope." Environmental Health Perspectives, 110 (Supp. 2): 183-193.

46. Maantay, J.A. (2001). Zoning, Equity, and Public Health. American Journal of Public Health, 91 (7):1033–1041.

47. Maantay, J.A. (2002). Mapping Environmental Injustices: Pitfalls and Potential of Geographic Information Systems (GIS) in Assessing Environmental Health and Equity. Environmental Health Perspectives, 110 (Supp. 2):161-171.

48. Maantay, J.A., and Maroko, A.R. (2009). Mapping urban risk: Flood hazards, race, & environmental justice in New York. Applied Geography, 29:111-124. Maantay, J.A., Maroko, A. R., and Culp, G. (2010). Using Geographic Information Science to Estimate Vulnerable Urban Populations for Flood Hazard and Risk Assessment in New York City, in Showalter, P., and Lu, Y. eds., Geotechnical Contributions to Urban Hazard and Disaster Analysis, Springer-Verlag:71-97.

49. Maas, J., Verheij, R. A., Groenewegen, P. P., de Vries, S., & Spreeuwenberg, P. (2006). Green space, urbanity, and health: How strong is the relation? Journal of Epidemiology and Community Health, 60(7): 587–592.

50. MacGregor, A.S. (1967). Public Health in Glasgow, 1905-1946. Edinburgh: E&S Livingstone.

51. MacIntyre, S., Ellaway, A., and Cummins, S. (2002). Place effects on health: how can we conceptualise, operationalise and measure them? Social Science & Medicine, 55:125–139.

52. MacIntyre, S. (2007). Inequalities in health in Scotland: What they are and what can we do about them? Glasgow: Medical Research Council, Social and Public Health Sciences Research Unit.

53. MacKeen, D. (2011). Making a Way Out of No Way. From Glasgow to Detroit: Urban Gardening and Community Building. 20 pages. Glasgow: Discussion paper of the Peter Gibson Memorial Fund.

54. Malik, S., Schecter, A., Caughy, M., Fixler, D.E. (2004). Effect of proximity to hazardous waste sites on the development of congenital heart disease. Archives of Environmental Health, 59(4):177-181.

55. Maroko, A.R., Maantay, J.A. Sohler, N.L., Grady, K.L., and Arno, P.S. (2009). The complexities of measuring access to parks and physical activity sites in New York City: a quantitative and qualitative approach. International Journal of Health Geographics, 8(1):34-56.

56. Mitchell, J.K., ed. (1999). Crucible of Hazard: Mega-Cities and Disasters in Transition. Tokyo: United Nations University Press.

57. Mitchell, R., Fowkes,.G., Blane, D., and Bartley, M. (2005). High rates of ischaemic heart disease in Scotland are not explained by conventional risk factors. Journal of Epidemiology and Community Health, 59:565-567.

58. Mitchell, R. and Popham, F. (2007). Greenspace, urbanity and health: relationships in England. Journal of Epidemiology and Community Health, 61: 681 – 683.

59. Mitchell, R. and Popham, F. (2008). Effect of exposure to natural environment on health inequalities: an observational population study. Lancet, 372(9650): 1655-1660.

60. Miyake, K., Maroko, A.R., Grady, K., Maantay, J.A., Arno, P.S. (2010). Not Just a Walk in the Park: Methodological Improvements for Determining Environmental Justice Implications of Park Access in New York City for the Promotion of Physical Activity. Cities and the Environment, 3(1): Article 8.

61. Neumann, C.M., Forman, D.L., Rothlein, J.E. (1998). Hazard screening of chemical releases and environmental equity analysis of populations proximate to toxic release inventory facilities in Oregon. Environmental Health Perspectives, 106(4):217–226.

62. New York City Mayor's Office of Environmental Coordination (2001). City Environmental Quality Review (CEQR) Technical Manual.

63. O'Neill, M.S., Jerrett, M., Kawachi, I., Levy, J.I., Cohen, A.J., Gouveia, N., Wilkinson, P., Fletcher, T., Cifuentes, L., and Schwarz, J. (2003).Health, wealth, and air pollution: advancing theory and methods. Environmental Health Perspectives, 111(16):1861-1870.

64. Ottmann, M., Maantay, J., and Grady, K. (2010). Urban Agriculture, Green Infrastructure, and Urban Ecology: A Case Study of the South Bronx, NYC. Cities and the Environment, 3(1): article 20.

65. Ottmann, M., Maantay, J. A.; Grady, K., and Fonte, N. (2012). Characterization of Urban Agricultural Practices and Gardeners' Perceptions in Bronx Community Gardens, New York City. Cities and the Environment, 5(1): Article 13.

66. Pacione, M. (1995). Glasgow: The Socio-Spatial Development of the City. Chichester: Wiley

67. Pearce, J., Richardson, E., Mitchell R., Shortt, N. (2010). Environmental Justice and Health: The Implications of the Socio-Spatial Distribution of Multiple Environmental Deprivation for Health Inequalities in the United Kingdom. Transactions of the Institute of British Geographers, 35:522-539.

68. Perry, R. (1844). Facts and Observations on the Sanitary State of Glasgow, Shewing the Connections Existing Between Poverty, Disease, and Crime. Glasgow: Gartnaval Press.

69. Raffensperger, C., and Tickner, J. (1999). Public Health and the Environment: Implementing the Precautionary Principle. Washington DC: Island Press.

70. Roemmich, J.N., L.H. Epstein, S. Raja, L. Yin, J. Robinson, and Winiewicz, D. (2006). Association of access to parks and recreational facilities with the physical activity of young children. Preventive Medicine, 43:437–441.

71. Rundle, A., Field, S., Park, Y., Freeman, L., Weiss, C.C., and Neckerman, K. (2008). Personal and neighborhood socioeconomic status and indices of neighborhood walk-ability predict body mass index in New York City. Social Science & Medicine, 67(12):1951-1958.

72. Russell, J. B. (1888). Life in One Room: Considerations for the Citizens of Glasgow. Glasgow: James MacLehose & Son.

73. Scandrett, E. (2007). Environmental justice in Scotland: policy, pedagogy and praxis. Environmental Research Letters, 2:1-7.
74. Scandrett, E. (2010). Environmental Justice in Scotland: Incorporation and Conflict. Chapter 5 in: Davidson, N., McCafferty, P. and Miller, D. (eds.) NeoLiberal Scotland: Class and Society in a Stateless Nation. Pp. 183-201. Cambridge: Cambridge Scholars Publishing.
75. Schmidtlein, M.C., Deutsch, R.C., and Piegorsch, W.W. (2008). A Sensitivity Analysis of the Social Vulnerability Index. Risk Analysis, 26 (4): 1099 – 1114.
76. Schukoske, J. (2000). Community Development Through Gardening: State and Local Policies Transforming Urban Open Space. Legislation and Public Policy, 3: 351-392.
77. Scottish Government (2008). Planning Advice Notice (PAN) 65 Planning and Open Space. Edinburgh: Scottish Government.
78. Scottish Government. (2009). Scottish Index of Multiple Deprivation, General Report and Technical Report. Edinburgh: A Scottish Government National Statistics Publication.
79. Scottish Government (2012). Vacant and Derelict Land Survey. Statistical Bulletin Planning Series. Edinburgh: A National Statistics Publication for Scotland.
80. Scottish Government (2010). Scottish Neighbourhood Statistics, online at http://www.sns.gov.uk/default.aspx. (last accessed on-line February 20, 2012).
81. The Scottish Government Directorate for Planning and Environmental Appeals (2012). Report to Scottish Ministers on the Glasgow and The Clyde Valley Strategic Development Plan.
82. Shaw, M., Thomas, B., Smith, G., and Dorling, D. (2008). The Grim Reaper's Road Map: An Atlas of Mortality in Britain. Bristol: Policy Press.
83. Sheppard, E., Leitner, H., McMaster, R.B., Hongguo, T. (1999). GIS based measures of environmental equity: exploring their sensitivity and significance. Journal of Exposure and Analytical Environmental Epidemiology, 9:18–28.
84. Shortt, N., Richardson, E., Mitchell, R., and Pearce, J. (2011). Re-engaging with the Physical Environment: A Health-Related Environmental Classification of the U.K., Area, Royal Geographical Society, 43: 76-87.
85. Smith, M., Moseley, D., Chetcuti, J., de Ioanni, M. (2008). Glasgow and Clyde Valley Integrated Habitat Networks Report. Roslin: Scottish Wildlife Trust and Forestry Commission, and Glasgow and Clyde Valley Green Network Partnership.
86. Smith, N., and DeFilippis, J. (1999). The Reassertion of Economics: 1990s Gentrification in the Lower East Side. International Journal of Urban and Regional Research, 23(4): 638-653.
87. Smyth, J.J. (2000). Labour in Glasgow 1896-1936: Socialism, Suffrage, Sectarianism. East Linton: Tuckwell Press.
88. Song, C. and Kulldorff, M. (2003). Power evaluation of disease clustering tests. International Journal of Health Geographics, 2(1):9.
89. Tate, E. (2012). Social Vulnerability Indices: A Comparative Assessment Using Uncertainty and Sensitivity Analysis. Natural Hazards, 63: 325-347.
90. Taulbut, M., Walsh, D., Parcell, S., Hanlon, P., Hartmann, A., Poirier, G., and Strniskova, D. (2009). The Aftershock of Deindustrialisation – trends in mortality in Scotland and other parts of post-industrial Europe. Glasgow: Glasgow Centre for Population Health.
91. Tobler, W. (1970) A computer movie simulating urban growth in the Detroit region. Economic Geography, 46(2): 234-240.

92. United Church of Christ's Commission for Racial Justice. (1987). Toxic Wastes and Race in the United States: A National Report on the Racial and SocioEconomic Characteristics of Communities with Hazardous Waste Sites. New York: United Church of Christ.

93. United Kingdom DoE. (various publication dates, from 1995). Industry Profile Guidance Documents (various industries). From http://publications.environment-agency.gov.uk/ (last accessed April 29, 2012).

94. United States Centers for Disease Control and Prevention/Agency for Toxic Substances and Disease Registry (CDC/ATSDR) (2008). The CDC/ATSDR Public Health Vulnerability Mapping.

95. Vinceti, M., Rovesti, S., Bergomi, M. (2001). Risk of birth defects in a population exposed to environmental lead pollution. Sci Total Environ, 278(1-3):23-30.

96. Von Hassel, M. (2002). The struggle for Eden: Community Gardens in New York City. Connecticut: Bergin & Garvey.

97. Walsh, D., Bendel, N., Jones, R. and Hanlon, P. (2010). It's not 'just deprivation': why do equally deprived UK cities experience different health outcomes? Public Health, 124(9):487–495.

98. Wang, J. (2011). The health impacts of brownfields in Charlotte, NC: A spatial approach. In: Geospatial Analysis of Environmental Health, Maantay, J.A. and McLafferty, S. eds. Dordrecht, NL: Springer Verlag.

99. Werlen B. (1993). Society, Action and Space: An Alternative Human Geography. London: Routledge.

100. World Health Organization (WHO) Commission on Social Determinants of Health. (2008). Closing the Gap in a Generation: Health Equity Through Action on the Social Determinants of Health. Geneva: World Health Organization.

101. White, H.L. (1998). Race, class, and environmental hazards. In: Environmental Injustices, Political Struggles. Camacho D, ed. Durham, NC: Duke University Press.

102. 596 Acres organization website http://596acres.org/ (last accessed April 29, 2012).

Co-benefits of Designing Communities for Active Living: An Exploration of Literature

James F. Sallis, Chad Spoon, Nick Cavill,
Jessa K. Engelberg, Klaus Gebel, Mike Parker,
Christina M. Thornton, Debbie Lou,
Amanda L. Wilson, Carmen L. Cutter and
Ding Ding

2.1 INTRODUCTION

Physical inactivity accounts for 5 million deaths annually worldwide [1]. Most people are not sufficiently active, and physical activity is declining in many countries [2]. This is a global problem with the biggest burden in low and middle income countries [3]. Increasing physical activity is a priority of the United Nations through its non-communicable disease initiative [4].

Physical activity has been engineered out of people's lives through urban planning and transportation investments that favor travel by automobile, labor-saving devices at home and in the work place, and a proliferation of electronic entertainment options [5],[6]. Built environments are worthy of special attention because they can affect virtually all residents of a community for many decades. The United Nations [4], World Health Organization [7], national physical activity plans [8], U.S. Guide to Community Preventive Services [9],[10], U.S. Institute of Medicine [11], and other scientific groups worldwide [12],[13] have identified creating built environments and implementing policies that support active living as essential for increasing physical activity and improving health.

Many decisions affecting physical activity environments occur at the local government level. Though mayors, city council members, and other officials work every day to balance competing interests, they likely do not consider that environments supporting physical activity could produce additional benefits for their communities. For example, changing zoning codes to favor mixed use developments can enhance property values and reduce carbon emissions [14],[15]. Having parks in neighborhoods has been linked with physical and mental health benefits [16].

There is no resource that examines the wide range of potential co-benefits of communities designed to support active living, which can be called "activity-friendly environments". Therefore, the aim of the current study was to compile evidence about the relation of activity-friendly environmental attributes to multiple potential outcomes. The expectation was that several co-benefits would be documented, but negative effects were also included. The intent of the present exploration of the literature was to provide an initial summary of the evidence on co-benefits that may be useful for determining whether and which systematic reviews are justified, identifying areas for further research, and educating policy makers about likely co-benefits.

2.2   METHODS

The present literature review covered diverse topics across multiple academic and practice fields, so dozens of systematic reviews were not feasible. Therefore an exploratory approach was taken to provide an initial overview of the potential co-benefits of communities designed for active living. We searched both scientific and gray literature. Gray literature was included because we expected many of the topic areas to be rarely addressed in the scientific literature but studied by government agencies and policy groups. The objective of the literature exploration was not to systematically review or quantify all evidence, but to create a profile of the potential multiple benefits, and negative effects, of each environmental feature as a tool for policy-making and a guide for future research.

2.2.1 Search areas

2.2.1.1 Built environment attributes

Specific built and social environment attributes in five settings (open spaces/parks/trails, urban design/land use, transportation, schools, workplaces/buildings) that research had shown to be related to physical activity were identified and used to structure the search (Table 1). The environmental attributes represented a multi-level conceptualization, based on ecological principles of multiple levels of influence on behavior and interactions across levels [17]. Proximity of activity-promoting settings was the most basic characteristic that could affect physical activity, such as proximity of parks or shops to homes. A second consideration was the quality or design of the setting, such as physical activity facilities within parks or quality of sidewalks. The third consideration was that social environments could interact with built environment features in affecting outcomes. For example, events and programs could improve use of well-designed parks, and social disorder like graffiti and boarded-up buildings could negate the benefits of well-designed streetscapes.

TABLE 2.1 Built and social environment features with evidence of association with physical activity.

Setting	Feature	Description	Reference
Open Spaces/ Parks/Trails	Design features	Size, amenities, physical activity facilities	[18-21]
	Presence/proximity	Existence of and distance to	[6,18,22]
	Trails	Proximity to and design of	[6,20]
	Programs, promotion, and events	Park-based programming	[19]
	Park incivilities/ civilities	Existence or lack of graffiti, litter, anti-social behavior (public drinking, loitering)	[20,23,24]
	Public gardens	Presence	[19]
Urban Design/ Land Use	Density	Population and housing density	[6,18,22]
	Mixed land use	Mix of destinations, distance to destinations	[6,18,22]
	Streetscale pedestrian design	Including buffers between street and sidewalk, building set-back from sidewalk, form based codes, street lights, etc.	[10,25]
	Greenery	Street trees/shrubbery, gardens	[18,22]
	Incivilities	Graffiti, vacant/dilapidated buildings, litter, anti-social behavior (public drinking, loitering)	[18,23,24]
	Accessibility & street connectivity	Density of intersections in street network	[18,20,26]

TABLE 2.1 *(Continued)*

Setting	Feature	Description	Reference
Transportation	Pedestrian/bicycle infrastructure	Sidewalks, bike lanes/paths, bike parking	[6,18,22,25]
	Crosswalk markings	Crosswalk and intersection quality	[25,27]
	Traffic calming	Speed bumps, curb-cuts, road diet, other engineering infrastructure	[27,28]
	Public transportation	Proximity to or density of bus, train stops	[6,20,22]
	Traffic speed/volume		[18]
	Safe routes to school	Engineering, programming, promotion and events	[6,21]
	Ciclovia/play streets	Opening streets for walking, bicycling, rolling, play	[26,29]
	Managed parking	Restricted parking access	[30]
Schools	School siting	Location of school, distance from residences (suburban, urban, rural)	[31]
	Recreation facilities	Physical education (PE) facilities and equipment, presence of PE teachers	[32,33]
	Shared use agreements	Community use of school facilities for physical activity	[33,34]
Buildings/ Workplaces	Building siting	Distance to residences, accessibility by public transit	[35]
	Mixed land use around worksite	Mix of destinations, distance to destinations	[6,18,22]
	Building site design	Design of property that building sits upon with physical activity options	[35]
	Building design	Stair design, exercise equipment presence, shower/locker presence, skip-stop elevators	[35,36]
	Worksite physical activity policies and programs	Exercise classes, discounted gym membership, active transportation promotion policies, parking cash out programs, point-of-decision prompts	[37,38]
	Workplace furniture design	Sit-stand desks	[39]

2.2.1.2   Co-benefits/outcomes

Based on the input of authors and informants, the following co-benefit outcomes were included in the searches: physical health, mental health, safety/injury prevention, social benefits, economic benefits, and environmental sustainability focusing on carbon emissions and air pollution (Table 2). These outcomes were defined as "co-benefits" because they were expected benefits of activity-friendly environments in addition to increased physical activity.

TABLE 2.2   Outcomes of activity-supportive built and social environments examined in searches.

Outcome/co-benefit	Description
Physical health	Chronic diseases, obesity
Mental health	Depression, anxiety, well being, quality of life
Social benefits	Neighborhood/social cohesion, human capital
Environmental sustainability benefits	Carbon dioxide emissions, pollutants
Safety/Injury prevention	Crime, violence, injury, pedestrian/bicycle and car crashes
Economic benefits	Land value, governmental infrastructure costs, real estate profitability, productivity/job performance, health care costs, economic performance of cities

2.2.2   Search strategies

2.2.2.1 Snowball sampling to identify sources

Through the Active Living Research (ALR, www.activelivingresearch.org) network, 20 experts in various disciplines were contacted to nominate 1) groups/ organizations working on the built environment and non-physical activity co-benefits, 2) key reports and papers, both peer-reviewed and gray literature, 3) websites, 4) case studies of cities that have implemented activity-friendly built environment changes, 5) recommendations for other experts. Of the 20 experts contacted, 13 provided input.

2.2.2.2   Supplementary literature search

Authors conducted additional literature searches on environmental features and each co-benefit outcome to supplement the expert input. Literature searches were conducted November 2013 through February 2014 using combined search terms of environment features "and" co-benefit outcomes. Abstractors were instructed to use multiple synonyms for search terms because terms vary by discipline.

     Search engines included Scopus, PubMed, Google Scholar, ISI Web of Science, MEDLINE, PsychINFO, Academic Search Premier, ClimateArk, and Google. Searches specific to European studies and carbon emission outcomes were conducted by invited international experts to enhance coverage of these topics. Due to the breadth of the overall search and differences across topics, reviewers developed search protocols specific to the topic area, but some guidelines were provided. Abstractors were encouraged to search for specialized search engines in their assigned fields. The initial goal was to be inclusive in finding relevant sources of information. For scientific literature, reviewers were instructed to find systematic or non-systematic reviews first. If reviews were located, then the individual studies did not need to be searched, except for publications since the latest review. In cases where a review paper did not provide adequate specificity or quantification in the findings, selected primary studies from that review paper were abstracted to illustrate specific findings. If reviews were not located, then individual studies were searched. For gray literature, reports from credible organizations were targeted, from such groups as government agencies, academic centers, and selected advocacy groups. Newspapers, magazines, and blogs were not searched, except to identify citations of or links to more credible reports.

2.2.3   Data extraction

During the data extraction process, basic information on the built environment feature, co-benefit, study sample characteristics, study methods, and major findings were coded in tables specific to each of the five settings. Then the strength of each piece of evidence was graded based on the source, and the direction of each association was noted (Table 3). To simplify interpretation, " + " denotes

TABLE 2.3   Scoring methods for summarizing the evidence

Score	Type of evidence
4.5	Peer-reviewed, systematic review paper (including meta-analysis)
4	Peer-reviewed, non-systematic review paper (from scientific literature) or non-peer-reviewed review paper (from gray literature)
3.5	Any (singular) peer-reviewed study
3	Any (singular) non peer-reviewed study, such as a technical report from a government agency or academic center
2	Non-analytic studies (for example, case reports, case series, simulations) or advocacy report without a clear literature review
1	Expert opinion, formal consensus
Score	**Direction of association**
+	A favorable association was found between feature and co-benefit (feature was associated with "better" level of co-benefit
-	An unfavorable association was found between feature and co-benefit (feature was associated with "worse" level of co-benefit
0 (zero)	No association or inconsistent evidence was found between feature and co-benefit

that a physical activity-promoting environmental feature was associated with a co-benefit in a "favorable" direction. For example, having parks nearby was associated with better mental health or fewer carbon emissions. Similarly, "-" denoted that a physical activity-promoting feature of the environment was inversely associated with a co-benefit. For example, higher residential density was associated with more air pollution. A code "0" represented lack of significant association in either direction or inconsistent findings. Due to the number and diverse types of studies from different fields, it was not possible to grade the quality of each study, as is done in systematic reviews. Extraction tables were cross-checked by other staff for accuracy and clarity.

2.2.4 Synthesizing the findings

To illustrate areas with strong evidence as well as research gaps, a matrix was created for each of the five settings that summarized evidence of associations between built environment features and co-benefits. Using a quasi-quantitative approach, results were summarized by summing the weighted evidence from each resource. Briefly, each piece of evidence was scored based on the source and type of study/report (i.e., "weights") and the direction of association. The weighted scores for associations in each direction were summed for each direction of effect category (" + ", "-", "0"), and recorded in the online tables. Thus, there was a weighted score for each direction, such as 24 " + ", 7 "-", and 4 "0". "Net" scores were calculated by subtracting the weighted negative and zero scores from the weighted positive scores. In the preceding example, 24 minus (7+4) equals 13 (summary score). To be conservative, negative and zero findings were subtracted from positive findings, so the summary scores roughly indicated both the quantity and quality of the evidence. To make the summarization process even more conservative, only resources with quality scores of 3 or above were included, as resources with a lower score, such as unreferenced advocacy documents and consensus reports, lacked credibility.

   Cells in each summary table were labeled based on summary scores. Table 4 presents the summary labels. A net score of 15 or above was considered strong evidence ([+++] or [---]), as this was equivalent to more than three systematic reviews consistently supporting the association between an environmental attribute and a co-benefit. Scores of 10–14 ([++] or [--]), and 4–9 ([+] or [-]), indicated good and moderate evidence, respectively. "Good" scores were equivalent to more than two reviews, and "moderate" scores were equivalent to at least one non-systematic review. Finally, a net score of less than 4 was considered insufficient evidence and was not labeled.

TABLE 2.4   Summary of scores and color codes for each level of evidence

Level of evidence	Range of scores	Code
Strong evidence of positive effect	15 and above (+)	[+++]
Good evidence of positive effect	10-14 (+)	[++]
Moderate evidence of positive effect	4-9 (+)	[+]
Insufficient evidence	3.5 (−) to 3.5 (+)	[0]
Moderate evidence of negative or null effect	4-9 (−)	[−]
Good evidence of negative or null effect	10-14 (−)	[−−]
Strong evidence of negative or null effect	15 and above (−)	[−−−]

2.3   RESULTS

Abstractors identified a total of 521 results from 221 sources, coming from at least 17 countries. Four hundred eighteen of these results were from a source with a quality score of at least "3" and were used in the results reported here. The five setting-specific tables with notes about each study and codes for findings are available online (Additional file 1).

2.3.1   *Open spaces/parks/trails*

Table 5 summarizes 69 findings from studies in the open space/parks/trails setting. Of the 36 cells representing attribute by co-benefit combinations in the table, 3 had strong evidence of co-benefits, 3 had good evidence, and 7 had moderate evidence. There was good to strong evidence for the association between park presence/proximity and all co-benefits, except for economic benefits. Moderate evidence supported that physical activity promotion programs in parks and open spaces were associated with four co-benefits of mental health, social benefits, environmental benefits, and safety/injury prevention. Public gardens had moderate evidence of social and safety/injury prevention benefits. There was good evidence that trails had economic benefits. Overall, there were 23 "blank cells" out of 36, indicating no or insufficient evidence.

TABLE 2.5 Open spaces/parks/trails summary scores

Built environment attribute	Physical health	Mental health	Social benefits	Environmental sustainability	Safety/injury prevention	Economic benefits
Presence, proximity	[+++] 54+ 3.5(0)	[+++] 88.5+	[+++] 26.5+4(0)	[++] 16+4(0)	[++] 11+	[0] 7.5+4(0)
Design features	[0] 3.5+		[+] 7.5+			
Trails						[++] 11.5+
Physical activity programs/promotion		[+] 4.5+	[+] 4	[+] 4+	[+] 4+	
Incivilities					[0] 3.5+	
Public gardens			[+] 4.5+		[+] 4.5+	

2.3.2   Urban design

There were 202 findings used in the summary of the urban design setting. Of 30 cells (Table 6), 8 had strong evidence of co-benefits, 5 had good evidence, and 6 had moderate evidence. In the urban design setting, 4 cells had moderate or good evidence of negative effects and one cell had strong evidence for negative effects, which was mixed use and safety/injury prevention. Mixed use, greenery, street scale design, and accessibility and street connectivity had evidence of 4 to 5 co-benefits. With the exception of street-scale design, urban design features had strong evidence of environmental benefits. All urban design features had evidence of economic benefits, and the evidence was particularly strong for mixed use. Of all urban design features, only greenery had strong evidence of mental health benefits. None had evidence of safety/injury prevention benefits. Residential density had the most complex pattern, with good evidence of negative health effects, strong evidence of environmental sustainability benefits, and good evidence of economic benefits.

2.3.3   Transportation systems

There were 81 findings in the transportation systems category. Of 48 cells (Table 7), 5 had strong evidence of co-benefits, 2 had good evidence, and 6 had moderate evidence. Strong evidence of co-benefits was most apparent in the safety/injury prevention and economic domains. Pedestrian and bicycle facilities had the best evidence of multiple co-benefits, followed by lower traffic speed and volume. Public transport had strong evidence of economic benefits and mixed evidence of environmental sustainability benefits. Overall, 34 of 48 combinations of environmental feature and co-benefit had no or inadequate evidence, showing many research gaps.

2.3.4   Schools

There were 27 findings in the school setting category. Of the 18 cells in Table 8, two cells had strong evidence of co-benefits, one had good evidence, and five had moderate evidence. Siting schools near the homes of students had strong evidence of environmental sustainability benefits and moderate evidence of mental health and economic benefits. Recreation facilities at schools and shared use agreements had evidence of multiple co-benefits.

TABLE 2.6 Urban design summary scores

Built environment attribute	Physical health	Mental health	Social benefits	Environmental sustainability	Safety/injury prevention	Economic benefits
Residential density	[--] 19+21.5(0) 7.5-		[-] 13.5+14.5(0)	[+++] 88+21(0) 3.5-	[--] 4.5(0) 7.5-	[++] 15+3.5(0)
Mixed land use	[+] 28+17(0) 4-	[0] 4.5+4	[+++] 33+11(0)	[+++] 95+21(0)	[---] 4.5(0) 11-	[+++] 22.5+3.5(0) 4-
Streetscale pedestrian design	[+] 7.5+		[+] 7.5+	[+] 7.5+		[+] 7+
Greenery	[+++] 20.5+3.5(0)	[+++] 26.5+	[++] 12+	[+++] 39.5+		[++] 12+
Accessibility & street connectivity	[++] 30+12(0) 7.5-		[++] 14.5+3.5(0)	[+++] 35.5+3.5(0)	[-] 4.5(0)	[+] 12.5+3.5(0)

TABLE 2.7 Transportation systems summary scores

Built environment attribute	Physical health	Mental health	Social benefits	Environmental sustainability	Safety/injury prevention	Economic benefits
Pedestrian/bicycle facilities		[0] 3+	[+] 7+	[+] 10.5 + 3.5(0)	[+++] 27.5 + 4(0)	[+++] 22.5 + 3.5(0)
Crosswalk markings					[--] 6(0) 4-	
Traffic calming	[0] 3.5+	[0] 3.5(0)	[0] 3+	[0] 3 + 3-	[+++] 23+	[0] 3+
Public Transportation	[0] 3.5-			[++] 28.5 + 17.5(0)		[+++] 20 + 4-
Traffic speed/volume	[0] 3.5+		[0] 3+	[+++] 14+	[+] 7+	[+] 7+
Safe routes to school			[0] 3+	[0] 3.5+	[+] 9.5 + 4(0)	
Ciclovia/play streets			[+] 7+			[0] 3.5+
Managed parking				[++] 10.5+		

TABLE 2.8 Schools summary scores

Built environ- ment attribute	Physical health	Mental health	Social benefits	Environ- mental sus- tainability	Safety/ injury pre- vention	Economic benefits
School siting	[0] 3.5+	[+] 4.5+		[+++] 21.5+	[0] 3-	[+] 4+
Recreation facilities	[++] 16+3.5(0)	[+++] 16.5+	[0] 3.5+			[0] 3.5+
Shared use agreements			[+] 7.5+		[+] 4+	[+] 7.5+

2.3.5 Workplaces/buildings

There were 39 findings in the workplace/building category. Of the 36 cells in Table 9, three cells had good evidence of co-benefits and three had strong evidence. Specifically, building site design (mainly outdoor) features had strong evidence of physical and good evidence of mental health benefits, and features of the building design had strong evidence of physical health and good evidence of environmental sustainability and economic benefits. Physical activity programs and policies within workplaces had strong evidence of economic benefits. For workplace and building features, the best evidence was for physical health and economic benefits.

TABLE 2.9 Workplaces/buildings summary scores

Built environ- ment attribute	Physical health	Mental health	Social benefits	Environ- mental sus- tainability	Safety/ injury pre- vention	Economic benefits
Building siting	[+] 4+					
Mixed land use around worksite				[+] 4+		[+] 4+
Building site design	[+++] 16+	[++] 11.5+				[0] 3.5+
Building design	[+++] 19.5+	[0] 3.5+4-		[++] 12.5+		[++] 12+
Worksite physical activity policies and programs	[+] 8.5+	[0] 3.5+		[+] 4+		[+++] 25+
Workplace furniture design	[0] 7+3.5(0)					[0] 3.5+3.5(0)

TABLE 2.10 Overall co-benefits by setting summary scores

Built environment attribute	Physical health	Mental health	Social benefits	Environmental sustainability	Safety/injury prevention	Economic benefits
Open spaces/Parks/Trails	[+++] 57.5+3.5(0)	[+++] 93+	[+++] 42.5+4(0)	[+++] 20+4(0)	[+++] 23+	[+++] 19+4(0)
Urban design	[+++] 105+54(0) 19-	[+++] 31+4	[+++] 80.5+29(0)	[+++] 265.5+45.5(0) 3.5-	[---] 13.5(0) 18.5-	[+++] 69+10.5(0) 4-
Transportation systems	[0] 7+3.5-	[0] 3+3.5(0)	[+++] 23+	[+++] 70+21(0) 3-	[+++] 67+14(0) 4	[+++] 56+3.5(0) 4
Schools	[+++] 19.5+3.5(0)	[+++] 21+	[++] 11+	[+++] 21.5+	[0] 4+3-	[+++] 15+
Workplaces/Buildings	[+++] 55+3.5(0)	[++] 18.5+4-		[+++] 20.5+		[+++] 48+3.5(0)

2.3.6   Overall summary of co-benefits by setting

In the final table, results were summed across features for each of the five settings. These results are intended to illustrate the overall potential for each setting to contribute to each co-benefit. Table 10 represents 418 findings. Of the 30 cells in the matrix, 22 had strong evidence of co-benefits and 2 had good evidence. Five cells had inadequate evidence, and only one cell had evidence of a net negative effect. Open spaces/parks/trails was the only setting with good to strong evidence of all six co-benefits. Activity-friendly design features in all five settings had strong evidence of environmental and economic benefits. Many gaps in the evidence existed for the transportation and workplace/buildings settings, particularly regarding the outcome of safety/injury prevention. There was little evidence of negative consequences of activity-friendly environments. However, in the urban design setting there was some evidence of negative physical health and safety/injury outcomes, mainly related to high residential density.

2.4   DISCUSSION

The present exploration of diverse peer-reviewed and gray literature revealed substantial documentation that designing communities that support physical activity for both recreation and transportation purposes is likely to produce a wide variety of additional benefits, ranging from mental health to environmental sustainability and economics. The present paper is the first attempt to compile such a wide range of evidence, and the results supported multiple potential co-benefits of designing environments for active living. This initial synthesis can guide future research and can serve as an interim tool for evidence-based decision-making regarding planning and design of built environments. A longer report version of the present study, with additional detailed information about methods, findings and additional sections on disparities and policy implications is presented online (Additional file 2).

When the results from all features were combined, there was impressive evidence of co-benefits in all settings, with 22 of 30 cells having strong evidence. For all settings, there was strong evidence for at least three of the six co-benefits of activity-friendly design. Within each setting there were several features that could be designed to create activity-friendliness. Thus, the present paper can be viewed as a menu of options that would allow designers and planners to devise multiple combinations of features to achieve activity-friendly environments. If there was strong evidence that a feature in a setting was related to the co-benefit, then designing this feature to support physical activity might yield the co-benefits indicated by this review.

There is mounting evidence that multiple environmental features or patterns of features combine to produce stronger effects on physical activity than any single feature [25],[40],[41], and this principle may apply to the co-benefits as well. If several features of a setting had relatively weak evidence of co-benefits, it is reasonable to expect that optimizing these multiple features could, in aggregate, produce strong effects. Thus, it is justified to sum the co-benefits scores across multiple features to estimate the potential overall effect of designing a setting to optimize physical activity, as was done to create Table 10.

The most studied setting was urban design, with more findings reviewed than all other settings combined. All features examined, including high residential density, mixed land use, activity-friendly street scale design, greenery, and high accessibility and street connectivity, were associated with environmental and economic benefits. In contrast, much less evidence was identified for the schools and workplaces/buildings settings. Blank cells in the summary tables indicate gaps to be filled by future research. For example, programs such as "Safe Routes to School" and Ciclovias or Open Streets may have numerous benefits, such as improved social benefits, reduced carbon emissions, and cleaner air, but these outcomes remain to be documented.

One notable finding was that economic benefits of activity-friendly designs were documented for all five physical activity settings. Based on the specific studies identified, many groups could enjoy economic benefits of activity-friendly environments, including governments (due to reduced spending on infrastructure), homeowners, real estate developers, health insurance companies, employers, retailers, commercial property owners, and taxpayers. This is an extremely broad range of beneficiaries, and some of them may not be aware of the economic benefits of activity-friendly environments.

2.4.1 Policy implications

Policy-makers worldwide are faced with many problems and challenges [42]. Rates of chronic disease and related costs are high in countries at all income levels, and these rates are increasing fastest in low- and middle-income countries [4]. Depression creates the highest burden of disease worldwide [43], and injuries are the biggest cause of death among young people [43]. The consequences of climate change are expected to be the worst human-made disasters in history [44]. Every country and city is looking for ways to improve economic growth. It seems inconceivable that making cities better for physical activity could contribute to solutions of all these problems. However, the evidence compiled here suggests designing activity-friendly communities could be a partial solution for many critical problems.

TABLE 2.11 Best evidence of environmental features with strong multiple benefits (at least "moderate" evidence of three benefits)

Setting	Built environment attribute	Evidence
Open Spaces/ Parks/Trails	Park presence/proximity	3 strong, 2 good
	Programs, promotion, and events	4 moderate
Urban Design/Land Use	Mixed land use	3 strong, 1 moderate (1 strong negative)
	Greenery	3 strong, 2 good
	Streetscale pedestrian design	4 moderate
	Accessibility and street connectivity	1 strong, 2 good, 1 moderate (1 good evidence of negative)
Transportation	Pedestrian/bicycle infrastructure	2 strong, 2 moderate
	Reduced traffic speed and volume	1 strong, 2 moderate
Schools	School siting	1 strong, 2 moderate
	Shared use agreements	3 moderate
Buildings/ Workplaces	Building design	1 strong, 2 good
	Physical activity policies and programs	1 strong, 2 good

If decisions about built environments were informed by evidence, then the features with the best evidence of co-benefits listed in Table 11 would deserve special consideration. The features listed in Table 11 had at least "moderate" evidence of three co-benefits. Among these, the best supported environmental features with at least "good" evidence of three co-benefits were park proximity, mixed land use, greenery, accessibility and street connectivity, building design, and workplace physical activity policies/programs.

2.4.2 Strengths and limitations

The strength of the literature exploration was the breadth of topics explored. For each setting several features were identified that were related to physical activity, and each of these features was evaluated for six types of co-benefits. A quasi-quantitative approach was used to code the level of evidence of 521 findings and weight each finding by the quality of the source. To avoid basing findings on lower quality evidence, such as poorly-substantiated claims in advocacy documents, lower-quality evidence was not included in the numerical summaries.

The main limitations were a consequence of the breadth. Because of the large number of topics searched, it was not possible to conduct a systematic review. A requirement of systematic reviews is assessment of the quality of each study, but this was not feasible given the number and diverse types of studies. It would have been helpful to code studies as being cross-sectional, longitudinal, or experimental in design. Existing reviews were used whenever possible to reflect the best evidence in the literature. Literature searches and coding were conducted by several investigators with a semi-structured process. Therefore there were undoubtedly differences across topics in thoroughness of search and classification of levels of evidence.

The searches were limited to English language documents, but one-quarter of the findings were from countries other than the US, and another quarter of findings were from reviews that included international literature. Another limitation was publication bias that favors positive findings, though this may have been countered somewhat by inclusion of gray literature, including technical reports. The intent of the literature exploration was to identify as many relevant sources as possible to determine whether it is worthwhile to pursue the topic of co-benefits, but a weakness was that the quality of each source was merely categorized and not based on an analysis of methodological quality.

The summary scores are not intended to be interpreted literally as the actual strength of evidence, but they provide a rough indication of the extent of evidence: pro, con, and neutral. If the net scores for the level of evidence are strong, there is reason to have confidence in the finding for a connection between a feature and an outcome because a strong rating required findings from multiple sources.

2.5 CONCLUSIONS

Substantial evidence indicated that designing and creating parks, communities, transportation systems, schools, and buildings that make physical activity attractive and convenient is also likely to produce a wide range of additional benefits. Present findings provide new information to decision-makers in numerous sectors that could change the perceived benefits of activity-friendly designs. Benefits were found for environmental sustainability, economics, and multiple dimensions of health. Though the present review is not definitive, the large number of sources identified for such policy-relevant issues provides a compelling justification for more original research and systematic reviews of the very broad range of topics. If "a good solution solves multiple problems," then building places that support physical activity may be considered a superlative solution.

REFERENCES

1. Lee IM, Shiroma EJ, Lobelo F, Puska P, Blair SN, Katzmarzyk PT. Effect of physical inactivity on major non-communicable diseases worldwide: an analysis of burden of disease and life expectancy. Lancet. 2012; 380(9838):219-29. doi:10. 1016/S0140-6736(12)61031-9
2. Hallal PC, Andersen LB, Bull FC, Guthold R, Haskell WL, Ekelund U. Global physical activity levels: surveillance progress, pitfalls, and prospects. Lancet. 2012; 380(9838):247-57. doi:10. 1016/S0140-6736(12)60646-1
3. Ng SW, Popkin BM. Time use and physical activity: a shift away from movement across the globe. Obe Rev. 2012; 13(8):659-80. Publisher Full Text OpenURL
4. United Nations. Political declaration of the high level meeting of the General Assembly on the prevention and control of non-communicable diseases (Sixth seventh session); 2011 [http://www.un.org/ga/search/view_doc.asp?symbol=A/66/L.1]
5. Brownson RC, Boehmer TK, Luke DA. Declining rates of physical activity in the United States: what are the contributors? Annu Rev Public Health. 2005; 26:421-43.
6. Sallis JF, Floyd MF, Rodriguez DA, Saelens BE. Role of built environments in physical activity, obesity, and cardiovascular disease. Circulation. 2012; 125:729-37.
7. World Health Organization. Global strategy on diet, physical activity and health. Geneva, Switzerland: World Health Organization; 2004 [http://www.who.int/dietphysicalactivity/strategy/eb11344/strategy_english_web.pdf?ua=1]
8. Bornstein DB, Pate RR, Pratt M. A review of the national physical activity plans of six countries. J Phys Act Health. 2009; 6 Suppl 2:S245-64. OpenURL
9. Centers for Disease Control and Prevention. Strategies to prevent obesity and other chronic diseases: the CDC guide to strategies to increase physical activity in the community. Atlanta, GA: U.S. Department of Health and Human Services; 2011 [http://www.cdc.gov/obesity/downloads/pa_2011_web.pdf]
10. Heath GW, Brownson RC, Kruger J, Miles R, Powell KE, Ramsey LT et al.. The effectiveness of urban design and land use and transport policies and practices to increase physical activity: a systematic review. J Phys Act Health. 2006; 3 Suppl 1:S55-76.
11. Koplan JP, Liverman CT, Kraak VI. Preventing childhood obesity: health in the balance. National Academies Press, Washington, DC; 2005.
12. Giles-Corti B, Foster S, Shilton T, Falconer R. The co-benefits for health of investing in active transportation. N S W Public Health Bull. 2010; 21:122-7.
13. National Institute for Health and Clinical Excellence. Promoting or creating built or natural environments that encourage and support physical activity. London: National

Institute for Health and Clinical Excellence; 2008 [http://www.pedestrians-int.org/content/45/72008_ph.pdf]

14. Frank LD, Sallis JF, Conway TL, Chapman JE, Saelens BE, Bachman W. Many pathways from land use to health: associations between neighborhood walkability and active transportation, body mass index, and air quality. J Am Plann Assoc. 2006; 72:75-87.

15. Frank LD, Greenwald MJ, Winkelman S, Chapman JE, Kavage S. Carbonless footprints: promoting health and climate stabilization through active transportation. Prev Med. 2010; 50 Suppl 1:S99-105.

16. Di Nardo F, Saulle R, La Torre G. Green areas and health outcomes: a systematic review of the scientific literature. Ital J Public Health. 2010; 7(4):402-13.

17. Sallis JF, Owen N, Fisher EB. Ecological models of health behavior. In: Health Behavior and Health Education: Theory, Research, and Practice. 4th ed. Glanz K, Rimer BK, Viswanath K, editors. Jossey-Bass, San Francisco, CA; 2008: p.465-86.

18. Ding D, Sallis JF, Kerr JK, Lee S, Rosenberg DE. Neighborhood environment and physical activity among youth: a review. Am J Prev Med. 2011; 41(4):442-55.

19. Godbey G, Mowen A, Ashburn VA. The benefits of physical activity provided by park and recreation services: the scientific evidence. National Recreation and Park Association, Ashburn, VA; 2010.

20. Sallis JF, Adams MA, Ding D. Physical activity and the built environment. In: The Oxford Handbook of the Social Science of Obesity. Cawley J, editor. Oxford University Press, New York, NY; 2011: p.433-51.

21. Stewart O, Moudon AV, Claybrooke C. Multistate evaluation of safe routes to school programs. Am J Health Promot. 2014; 28(sp3):S89-96. Publisher Full Text

22. Bauman AE, Reis RS, Sallis JF, Wells JC, Loos RJF, Martin BW. Correlates of physical activity: why are some people physically active and others not? Lancet. 2012; 380(9838):258-71. doi:10. 1016/S0140-6736(12)60735-1

23. Miles R. Neighborhood disorder, perceived safety, and readiness to encourage use of local playgrounds. Am J Prev Med. 2008; 34(4):275-81.

24. Molnar BE, Gortmaker SL, Bull FC, Buka SL. Unsafe to play? neighborhood disorder and lack of safety predict reduced physical activity among urban children and adolescents. Am J Health Promot. 2004; 18(5):378-86.

25. Cain KL, Millstein RA, Sallis JF, Conway TL, Gavand K, Frank LD et al.. Contribution of streetscape audits to explanation of physical activity in four age groups based on the microscale audit of pedestrian streetscapes (MAPS). Soc Sci Med. 2014; 116:82-92.

26. Sarmiento OL, Torres A, Jacoby E, Pratt M, Schmid TL, Stierling G. The ciclovia-recreativa: a mass-recreational program with public health potential. J Phys Act Health. 2010; 7 Suppl 2:S163-80.

27. Badland H, Schofield G. Transport, urban design, and physical activity: an evidence-based update. Transport Res D-TR E. 2005; 10(3):177-96.

28. Rothman L, Buliung R, Macarthur C, To T, Howard A. Walking and child pedestrian injury: a systematic review of built environment correlates of safe walking. Inj Prev. 2014; 20:41-9.

29. Torres A, Sarmiento OL, Stauber C, Zarama R. The ciclovia and cicloruta programs: promising interventions to promote physical activity and social capital in Bogota, Colombia. Am J Public Health. 2013; 103(2):e23-30.

30. Parking spaces/community spaces: finding the balance through smart growth solutions. U.S. Environmental Protection Agency, Washington, DC; 2006.

31. McMillan TE. Walking and biking to school, physical activity, and health outcomes. A research brief. Active Living Research, a National Program of the Robert Wood Johnson Foundation, Princeton, NJ; 2009. OpenURL

32. Davison KK, Lawson CT. Do attributes in the physical environment influence children's physical activity? a review of the literature. Int J Behav Nutr Phys Act. 2006; 3:19. doi:10.1186/1479-5868-3-19

33. Spengler JO. Promoting physical activity through the shared use of school and community recreational resources. A research brief. Active Living Research, a National Program of the Robert Wood Johnson Foundation, Princeton, NJ; 2012.

34. Lafleur M, Gonzalez E, Schwarte L, Banthia R, Kuo T, Verderber J et al. Increasing physical activity in under-resourced communities through school-based, joint-use agreements, Los Angeles County, 2010–2012. Prev Chronic Dis 2013, 10; http://dx.doi.org/10.5888/pcd10.120270.

35. Zimring C, Joseph A, Nicoll GL, Tsepas S. Influences of building design and site design on physical activity: research and intervention opportunities. Am J Prev Med. 2005; 28(2Suppl 2):186-93.

36. Nicoll G, Zimring C. Effect of innovative building design on physical activity. J Public Health Policy. 2009; 30 Suppl 1:S111-23.

37. Crespo NC, Sallis JF, Conway TL, Saelens BE, Frank LD. Worksite physical activity policies and environments in relation to employee physical activity. Am J Health Promot. 2011; 25(4):264-71.

38. Proper KI, Koning M, van der Beek AJ, Hildebrandt VH, Bosscher RJ, van Mechelen W. The effectiveness of worksite physical activity programs on physical activity, physical fitness, and health. Clin J Sport Med. 2003; 13(2):106-17.

39. Alkhajah TA, Reeves MM, Eakin EG, Winkler EAH, Owen N, Healy GN. Sit-stand workstations: a pilot intervention to reduce office sitting time. Am J Prev Med. 2012; 43(3):298-303.

40. Adams MA, Sallis JF, Kerr J, Conway TL, Saelens BE, Frank LD et al.. Neighborhood environment profiles related to physical activity and weight status: a latent profile analysis. Prev Med. 2011; 52:326-31.

41. Sallis JF, Bowles HR, Bauman A, Ainsworth BE, Bull FC, Craig CL et al.. Neighborhood environments and physical activity among adults in 11 countries. Am J Prev Med. 2009; 36(6):484-90.

42. Frank LD, Saelens BE, Powell KE, Chapman JE. Stepping toward causation: do built environments or neighborhood and travel preferences explain physical activity, driving, and obesity? Soc Sci Med. 2007; 65(9):1898-914.

43. Global health risks: mortality and burden of disease attributable to selected major risks. WHO Press, Geneva, Switzerland; 2009.

44. Climate change 2014: impacts, adaptation, and vulnerability (IPCC working group II contribution to fifth assessment report). Cambridge University Press, Cambridge, United Kingdom and New York, NY; 2014.

Supplemental material available online at http://www.ijbnpa.org/content/12/1/30.

CHAPTER 3

Why We Need Urban Health Equity Indicators: Integrating Science, Policy, and Community

Laura M. Jason Corburn and Alison K. Cohen

3.1 TOWARD HEALTHIER AND MORE EQUITABLE CITIES

As the world urbanizes, global health challenges are increasingly concentrated in cities. Currently, over 80% of the population in Latin America already lives in cities. The African urban population is projected to double in the next decade and China has urbanized in thirty years at a rate it took Europe and North America a century [1]. Rapidly growing new cities and increasingly segregated older cities in the global north and south are contributing to health inequities. Urban planning and policies can influence population health by supporting or stymieing opportunities for employment, housing security, political participation, education, protection from environmental risks, access to primary health care, and a host of other social and physical determinants of well-being [2]. As the World Health Organization (WHO) and UN-HABITAT acknowledged in the 2010 report entitled "Hidden Cities: Unmasking and Overcoming Health Inequities in Urban Settings", where in a city you live and how that city is governed can determine whether or not one benefits from city living [3].

Measuring the forces that contribute to urban health is one challenge for promoting more healthy and equitable cities. Burden of disease estimates have tended to focus on the whole world or specific geographic regions [4],[5]. These data can mask intra-city differences and global data may not be relevant to inform national or municipal policy making. Public health has developed metrics for single pathogenic exposures or risk factors, but these measures often ignore both community assets that promote health equity and the cumulative impacts on health from exposure to multiple urban environmental, economic, and social stressors [6],[7]. Recognizing these population health challenges, the United Nations (UN) Commission on Social Determinants of Health (2008) called for "health equity to become a marker of good government performance" ([8], p. 11) and for the UN to "adopt health equity as a core global development goal and use a social determinants of health indicators framework to monitor progress" ([8], p. 19). More recently, the 2011 World Social Determinants of Health Conference and the Pan-American Health Organization's Urban Health Strategy called for the development of new urban health equity indicators that track the drivers of health inequities across place and time, particularly within a city neighborhood [9] (Box S1). In this paper, we briefly outline an approach for promoting greater urban health equity through the drafting and monitoring of indicators. We draw examples from the cities of Richmond, California, and Nairobi, Kenya. More specifically, we argue that participatory indicator processes hold the potential to shape new healthy and equitable urban governance by:

- integrating science with democratic decision making;
- tracking policy decisions that shape the distribution of health outcomes; and
- including protocols for ongoing monitoring and adjusting of measures over time.

3.2   INDICATORS FOR HEALTH PROMOTING POLICY CHANGE

Ongoing measurement and evaluation is one critical aspect of moving toward more healthy and equitable cities because what we measure often matters for whether and how we act. Yet, the danger of indicator efforts is that they portray a too simplified picture of a complex reality and policy solutions may suffer the same defects. For example, indicators of single chemical exposures cannot produce policy-relevant knowledge about the environmental health consequences of multiple exposures. In a similar way, cross-sectional measures of single built and social environmental features of urban neighborhoods tend to ignore the cascading and relational effects of inequalities in urban areas. For example, in our own work in the slums of Nairobi, we have found that typical indicators that only measure population access to a toilet can misconstrue whether an ablution block is hygienic or safe. In Nairobi's slums, accessing a toilet may be controlled by a local cartel that might extort a high price for users, disproportionately impacting family income, while at the same time acting as a location for rape and sexual violence against women, particularly at night, when the toilet has no lighting or security, which in-turn might contribute to the spread of sexually transmitted diseases. Capturing the relationships between a physical or economic measure, the political decisions that shape the distribution of community resources, and how urban residents currently navigate urban inequities to stay healthy is central to our understanding of effective and meaningful urban health equity indictors.

The WHO defines an indicator as a variable with characteristics of quality, quantity, and time used to measure, directly or indirectly, changes in health and health-related situations [10]. In this paper, we use the term indicator to represent a construct often consisting of more than one measure, with a metric being the actual quantitative or qualitative data that is used to populate the indicator [11]. Our review of the vast health equity indicator literature suggests that indicators should do more than capture health outcomes, but also the determinants

TABLE 3.1 Comparison of conventional and our approach to urban health equity indicators.

Characteristic	Conventional Health Status Indicator	Urban Health Equity Indicators
Time	Cross-sectional	Longitudinal: tracks progress over time
Orientation	Deficit-based	Asset-driven: strikes balance between identifying problems & building upon strategies that are already working
Levels	Individual behavior & biologic	Focuses on individual & community characteristics plus local, national, and international policies
Populations & places	Static	Dynamic: acknowledges that populations change and that definitions of community will change
Accessibility	Expert-driven	Collaborative & participatory between professionals & community members
Policy relevance	Health sector	Explicitly linked to policy making institutions within and outside the health sector
Political power	Unclear	Emphasizes accountability and transparency in political process and distribution of state resources

of health that often drive outcomes, including institutional practices and policy decisions made outside the health care and medical sectors. In addition, effective urban health equity indicators ought to highlight associations between determinants and health impacts, use data that are verifiable and easily accessible, and be shared in a clear and compelling way to a range of interested stakeholders [11],[12],[13],[14] (Table 1 and S1).

3.3 INDICATORS AS ADAPTIVE URBAN HEALTH EQUITY GOVERNANCE

The complexity of cities and the variegated forces that contribute to (in)equity in urban neighborhoods demands that indicator development processes are similarly dynamic. The drafting, measuring, tracking, and reporting of indicators can be viewed not as a technical process for experts alone, but rather as an opportunity to develop new participatory science policy making, or what we call governance. Governance is not just government and the decisions of formal institutions, such as ministries of health, but also includes the norms, routines, and practices that help shape which issues get onto the health research and policy agenda, what evidence base is used to underwrite decisions, and which social actors are deemed expert enough to participate in these decisions [15]. In other words, governance processes can shape what issues are deemed important for promoting health equity and which institutions are responsible for action [16] (Box 1).

BOX 1. HOW INDICATORS ACT AS A FORM OF HEALTHY URBAN GOVERNANCE

- Identifying and framing what counts as a health policy issue.
- Generating, or contributing to, the evidentiary standards that underwrite health equity issues.
- Constituting some social actors as "experts", by deciding who gets to participate in defining indicators.
- Grappling with different knowledge claims as the weight and importance of indicators is debated.
- Highlighting the importance of public accountability and transparency of data in the way indicators are reported and shared with various publics.

The UN-HABITAT Urban Indicators project recognizes the importance of governance measures for tracking urban equity. In order to measure governance, the UN project measures such things as the degree of decentralization in public decision making, voter participation, the number of participants in civil society organizations, and the public transparency and accountability of local government institutions [17]. Similarly, the World Health Organization's Urban Health Equity Assessment and Response Tool (HEART) also attempts to measure some aspects of governance related to health equity and includes indicators such as government spending on health and education, voter participation, percentage of population completing primary education, and the proportion of the population covered by health and other insurance [18]. These are important steps in acknowledging and capturing the role of politics and non–health care specific policy making in measuring and acting to promote greater health equity in cities.

We suggest that indicator processes themselves, not just the measures, can act as opportunities for crafting new healthy and equitable urban governance. While this is an emerging idea for city health management, ecologists and others have used an iterative governance process called adaptive management for decades to steward complex ecosystems, such as forests, wetlands, and fisheries [19]. Adaptive management acknowledges the failures of linear processes where narrow disciplinary scientists have aimed to develop complex models, predict long-term outcomes, and suggest one-time policy standards. Instead, adaptive management begins with an acknowledgement of the inherent complexity and uncertainty within systems, that this complexity demands an iterative, ongoing learning process among a range of expert stakeholders, and that policy interventions must be adjusted to reflect newly acquired knowledge [20]. Another difference between adaptive management and conventional science policy is that adaptive management does not postpone actions until definitive causality is known about a system, but rather emphasizes the importance of action in the face of uncertain science and couples decisions tightly to rigorous monitoring [21],[22].

The process of adaptive management is one where a broad group of stakeholders, from scientists to policy makers to users of a resource, work together to generate evidence, make decisions, monitor the progress of those decisions, and make ongoing adjustments to decisions as new information emerges from monitoring [20]. Gohlke and Portier call for greater capacity within public health institutions to adapt to new and emerging challenges, such as drug-resistant infections and climate change, and that the field and discipline is currently ill-suited for adaptive science-based research and practice [21]. Huang et al. also stress the importance of enhancing resources and training for the redesign of public health institutions to enhance the field's adaptive capacity [22]. Yet, few others in public health or urban planning have explored the potential of applying adaptive management to promote greater health equity in cities.

3.4   URBAN HEALTH INEQUITIES IN NORTH AND SOUTH

Drawing from our collaborative work on healthy urban governance and the drafting of health equity indicators in Richmond, California, and the Mathare Valley informal settlement in Nairobi, Kenya, we offer some brief examples of what an urban health equity adaptive management strategy might entail.

In Richmond, California, located in the San Francisco Bay Area, one-third of residents live at or below 200% of the federal poverty line, over 60% of the population is African-American, Latino, or Asian-American, one in seven people are unemployed, residents live with elevated concentrations of industrial and mobile source air pollution, and it is one of the most violent cities, measured by per-capita homicide rates, in the United States [23]. In Richmond, African-Americans have the highest rate of infant mortality, low-birth weight babies, and asthma hospitalizations in Contra Costa County, and residents of the Iron Triangle neighborhood, one of the poorest in the city, die on average 13 years earlier than their wealthier white neighbors [23].

The Mathare Valley is a sprawling informal settlement in Nairobi, where over 82% of residents rent dirt floor, sheet metal–walled, one-room shacks and 88% lack access to clean and reliable drinking water or a private, hygienic toilet [24] (Figure S1). According to a 2011 household survey conducted in Mathare, over two-thirds of residents experienced routine violence in the last year, over half live on environmentally risky slopes and flood-prone areas, and households spend, on average, 76% of their monthly income on food [24]. Child mortality in Nairobi's slums is 151 per 1,000 live births, compared to 62 per 1,000 in all of Nairobi and 113 in rural Kenya [25].

Recognizing these health inequities and the multiple factors contributing to them, actions to improve the physical, social, and economic environments are occurring in both Richmond and Mathare. In Richmond, community groups and the city government drafted a Health and Wellness Element—or a development and policy blueprint—as part of the city's General Plan Update. Community-based organizations also led their own processes to collect data for and draft health equity indicators [26],[27].

In Mathare, community groups are working to reduce violence, engage youth in employment activities, and build toilets and schools [28],[29]. One coalition includes Muungano wa Wanavijiji, the federation of the urban poor in Kenya, Slum Dwellers International (SDI), the University of Nairobi, and the University of California, Berkeley, who together are organizing residents to plan for physical and social improvements that include new water and sanitary infrastructure, housing and land rights, and environmental and health care services [30],[31]. The government of Kenya and the World Bank recently launched a national policy initiative aimed at improving living conditions and well-being in slums, called the Kenya Informal Settlements Improvement Programme (KISIP) [32].

3.5   URBAN HEALTH EQUITY INDICATORS IN PRACTICE

In Richmond, the indicator process emerged from ongoing community organizing and land use planning, and included community-based organizations and the city and county health department. Community priorities were highlighted through a process called "Measuring What Matters" where over ten different community-based organizations identified priority issues, chose indicators, collected and analyzed data, and published a comprehensive report that included quantitative and qualitative information [27]. At the same time, the city organized a participatory process to draft and implement the Health and Wellness Element, which included a set of goals and metrics aimed at promoting and monitoring progress on population health [33]. In order to track and monitor indicators on an ongoing basis, the Richmond Health Equity Partnership (http://richmondhealth.org) was established in 2012 and includes representatives from the city, county health department, school district, and a host of community-based organizations.

In Mathare, the nongovernmental organization Muungano Support Trust (MuST) has organized residents to survey themselves and document community assets and vulnerabilities in three waves starting in 2007 through 2012 [28],[31]. These data have been combined with spatial maps of community assets and hazards and used by MuST in community planning processes focused on specific projects, such as improving housing, water infrastructure, health care access, and community facilities [31]. In 2011, a comprehensive slum redevelopment plan focused on Mathare was drafted by residents, MuST, the University of Nairobi, and University of California, Berkeley, which includes indicators and a process for ongoing monitoring [31].

In both cases, we began by organizing community health priorities into three broad health equity categories: living conditions, economics and services, and political power and outcomes. Under each category, indicators were selected that constituted the elements of the category. For instance, under living conditions, housing, key utilities such as water, sanitation and food, the physical environment, community safety, and transportation were selected as representative indicator categories (Table 2).

Importantly for public health practitioners, policy makers, and community residents, each indicator includes a health- and equity-based rationale that is referenced in the peer-reviewed literature. While a number of measures could populate each indicator, the participatory process selected one or two priority measures that were deemed representative of the larger equity issue being addressed and linked these to local or state policies that were understood by participants as potentially promoting greater health equity. The idea was to generate a set of measures that when combined could suggest whether or not the community was making progress toward greater health equity (Figure S2).

3.6   LIMITS OF CITY HEALTH EQUITY INDICATORS

All indicator efforts are limited in that they make judgments about selecting and highlighting certain data over others. These value judgments do not make indicator efforts unscientific or invalid, but rather demand that the processes for selection and the denial of data be explicit and transparent and, as we have suggested here, open to interpretation and re-evaluation. Another limitation of indicator efforts is whether they inspire action by different actors, from within and outside the health sector. Traditional indicators that measure morbidity and mortality tend to either place responsibility for improving health on the medical and public health communities or on vaguely identified institutions such as the economy, education, or built environment. The result is an overemphasis on medical and public health solutions while failing to articulate the specific institutions and policies that might need to change to promote greater health equity. While our examples from Richmond and Nairobi are in the early stages, we are witnessing non–health care specific sectors and institutions, from urban planning to legal and housing rights to violence prevention, re-framing their work as contributing to urban health equity.

Indicator projects can also be limited by a lack of available data and the costs of obtaining locally specific information. Very few cities in the global north or south collect data on the social determinants of health at the neighborhood

TABLE 3.2   Examples of urban health equity indicators.

Equity Category	Indicators	Example Measures for Richmond, CA	Example Measures for Mathare Informal Settlement, Nairobi, Kenya
Living conditions	Housing	• Percentage of eligible residents receiving housing subsidies (i.e., Section 8) • Number of rehabilitated, formerly foreclosed/vacant housing properties	• Percentage residents in savings program for housing • Ratio of structure owners to tenants
	Water, sanitation, & food	• Ratio of eligible persons to number receiving food supports • Self-reports of food insecurity	• Self-reports of food insecurity • Percent of households with in-home water & toilet service • Number of new electricity connections installed by utility company
	Environment	• Percentage households reporting air pollution or noise-altered sleep, concentration, or work/school performance.	• Number of infrastructure projects launched to secure housing on steep slopes & in flood areas • Number of non-charcoal burning cook-stoves sold at subsidized cost
	Safety	• Perception of safety, especially at night • Percentage participating in community policing/cease-fire activities	• Self reports of safety & violence from women
	Transportation	• Public spending on bus and rail transport as ratio of highway spending	• Public spending on transport
Economics and services	Primary health care	• Percent of adults who did not seek medical care because of the cost • Number of new community health workers at clinics & other providers	• Percentage free clinics offering maternal and childhood care using in-home community health workers
	Mental/substance care	• Percentage county budget funding formerly incarcerated community members to receive counseling & care	• Percent of international health research budgets spent on mental health services/interventions
	Education	• Percentage subsidized enrollment in youth after-school programs	• Percent families receiving free day care

TABLE 3.2 (*Continued*)

Equity Category	Indicators	Example Measures for Richmond, CA	Example Measures for Mathare Informal Settlement, Nairobi, Kenya
	Employment	• Percent local employers offering living wage jobs, paid sick days, & health care/insurance	• Percent of local residents hired to work on government and internationally funded contracts in past year
	Wealth access	• Number of new business permits issued by the city • Amount of Community Reinvestment Act funds spent in city	• Ratio of slum dwellers' new bank accounts to all new accounts by local banks in past year
Political power & outcomes	Community participation	• Number of community members & local organization representatives elected and/or appointed to city and county boards & commissions	• Percentage of residents participating in community-based organization
	Government responsiveness	• Percentage of public works complaints responded to within 30 days or less • Percentage of public participation processes that are held at convenient times, and provide transportation & language translation	• Number of meetings held in community by Nairobi's city council and water & power company addressing ongoing infrastructure, housing, & health issues
	Recognition of minority rights (women)	• Percentage of residents reporting experiences of gender or ethnic discrimination in school, government relations, police interaction, and/or workplace	• Number of women given land rights/housing tenure by City Council
	Health status	• Self-rated health	• Self-rated health
	Art/cultural expression	• Per-capita funding for the arts	• Percentage of youth and adults participating in cultural programs

scale and those that do rarely keep these data in one publically accessible location. However, advances in mobile information technology and the use of hand-held devices with built-in sensors are creating new opportunities for tracking and reporting different types of urban health equity data. In Toronto, Canada, the health ministry has created the Toronto Central Local Health Integration Network (LHIN), which aims to bring together multiple community actors and government agencies to improve health for the urban poor and tracks progress using indicators of equity [34]. In Rio de Janeiro, Brazil, the Center for Health Promotion is a network of over 150 civil society organizations working to promote health equity and, among other tasks, gathers data on the social determinants of health equity in Rio's favelas [35]. In Belo Horizonte, Brazil, an urban health observatory conducts participatory research and maintains data on population- and place-based health equity issues [36]. The Indian nongovernmental organization Urban Health Resource Centre (http://www.uhrc.in) works with the urban poor to improve health equity and helped shape India's National Urban Health Mission, which will document many determinants of health in cites [37]. In San Francisco, California, the public health department maintains a health equity–oriented publically available database called the Healthy Development Measurement Tool [38].

3.7   CONCLUSIONS

As urban health is increasingly recognized as a global health priority, new indicators accompanied by monitoring processes that can adapt and improve over time will be necessary to promote greater health equity. We have suggested here that indicator processes might be one important strategy to encourage new models of urban health governance in both the global north and south. Like any concept, more research and evaluation is necessary to understand the barriers and opportunities for turning our conceptual ideas into practice. Yet, lessons from other fields and emerging experiments around the world suggest that indicator processes can integrate science, policy, and community to promote greater urban health equity.

REFERENCES

1. World Bank (2011) Urban development data. Available: http://data.worldbank.org/topic/urban-development. Accessed 1 December 2011.

2. Vlahov D, Freudenberg N, Proietti F, Ompad D, Quinn A, et al. (2007) Urban as a determinant of health. J Urban Health 84(1):16–25. doi: 10.1007/s11524-007-9169-3

3. World Health Organization/UN-HABITAT (2010) Hidden cities. Available: http://www.hiddencities.org/report.html. Accessed 8 July 2011.

4. Murray CJ, Lopez AD (1997) Mortality by cause for eight regions of the world: Global Burden of Disease Study. Lancet 349: 1269–1276. doi: 10.1016/s0140-6736(96)07493-4

5. Wurthwein R, Gbangou A, Sauerborn R, Schmidt CM (2001) Measuring the local burden of disease. A study of years of life lost in sub- Saharan Africa. Int J Epidemiol 30: 501–508. doi: 10.1093/ije/30.3.501

6. Steenland K, Armstrong B (2006) An overview of methods for calculating the burden of disease due to specific risk factors. Epidemiology 17: 512–519. doi: 10.1097/01.ede.0000229155.05644.43

7. Harpham T (2009) Urban health in developing countries: what do we know and where do we go? Health and Place 15: 107–116. doi: 10.1016/j.healthplace.2008.03.004

8. Commission on Social Determinants of Health (CSDH) (2008) Closing the gap in a generation: health equity through action on the social determinants of health. Final Report of the Commission on Social Determinants of Health. Geneva, World Health Organization doi: 10.1177/1757975909103709

9. World Health Organization (2011) Rio Political Declaration on Social Determinants of Health. World Conference on Social Determinants of Health. Available: http://www.who.int/sdhconference/declaration/en. Accessed 1 December 2011.

10. World Health Organization (2009) Indicator definitions. Available: http://www.who.int/whosis/indicators/en. Accessed 8 May 2011.

11. Flores LM, Davis R, Culross P (2007) Community health: a critical approach to addressing chronic diseases. Prev Chronic Disease 4(4): A108. Available: http://www.cdc.gov/pcd/issues/2007/oct/07_0080.htm. Accessed 14 June 2011.

12. O'Neill M, Simard P (2006) Choosing indicators to evaluate Healthy Cities projects: a political task? Health Promot Int 21(2):145–152. doi: 10.1093/heapro/dal006

13. Corburn J (2005) Street science: community knowledge and environmental health justice. Cambridge, MA: The MIT Press.

14. Cummins S, Curtis S, Diez-Roux AV, Macintyre S (2007) Understanding and representing 'place' in health research: a relational approach. Soc Sci Med 65: 1825–1838. doi: 10.1016/j.socscimed.2007.05.036

15. Burris S, Hancock T, Lin V, Herzog A (2007) Emerging strategies for healthy urban governance. J Urban Health 84(1):i54–i63. doi: 10.1007/s11524-007-9174-6

16. Corburn J (2009) Toward the healthy city: people, places, and the politics of urban planning. Cambridge, MA: The MIT Press.

17. UN-HABITAT (2011) Urban indicators. Available: http://www.unhabitat.org/content.asp?typeid=19&catid=646&cid=8383. Accessed 11 November 2011.

18. Word Health Organization (WHO) (2010) Urban Health Equity Assessment and Response Tool (HEART). Available: http://www.who.int/kobe_centre/measuring/urbanheart/en/index.html. Accessed 3 October 2011.

19. Lee KN (1993) Compass and gyroscope: integrating science and politics for the environment. Washington, D.C.: Island Press.

20. National Research Council (2004) Adaptive management in water resources project planning. Panel on Adaptive Management for Resource Stewardship. Washington, DC: National Academies Press.
21. Gohlke JM, Portier CJ (2007) The forest for the trees: a systems approach to human health research. Environ Health Perspect 115(9):1261–1263. doi: 10.1289/ehp.10373
22. Huang C, Vaneckova P, Wang X, FitzGerald G, Guo Y, Tong S (2011) Constraints and barriers to public health adaptation to climate change. Am J Prev Med 40(2):183–190. doi: 10.1016/j.amepre.2010.10.025
23. Contra Costa Health Services (CCHS) (2010) Community health indicators for Contra Costa County. Available: http://cchealth.org/health_data/hospital_council/. Accessed 23 February 2011.
24. Muungano Support Trust (MuST) (2011) Mathare Valley community survey. University of Nairobi and University of California, Berkeley. Unpublished data. Available: http://muunganosupporttrust.wordpress.com/ and http://www.centreforurbaninnovations.com/. Accessed 13 December 2011.
25. Kenya National Bureau of Statistics (KNBS) (2010) Demographic Health Survey (DHS), 2003–2009, results. Nairobi, Kenya
26. Cohen AK, Lopez A, Malloy N, Morello-Frosch R (2012) Our environment, our health: a community-based participatory environmental health survey in Richmond, CA. Health Educ Behav 39(2):198–209 doi:10.1177/1090198111412591. doi: 10.1177/1090198111412591
27. Pacific Institute (2009) Measuring what matters: neighborhood research for economic and environmental health and justice in Richmond, North Richmond and San Pablo. Available: http://www.pacinst.org/reports/measuring_what_matters/index.htm Accessed 8 June 2011.
28. Weru J (2004) Community federations and city upgrading: the work of Pamoja Trust and Muungano in Kenya. Environment and Urbanization 16(14):47–62. doi: 10.1177/095624780401600105
29. Karanja I (2010) An enumeration and mapping of informal settlements in Kisumu, Kenya, implemented by their inhabitants. Environment and Urbanization 22(1):217–239. doi: 10.1177/0956247809362642
30. Slum Dwellers International (SDI) (2011) What is incrementalism, part 2: community-managed utilities in an informal settlement in Nairobi. Available: http://www.sdinet.org/blog/2011/05/23/what-is-incrementalism-part-2-/. Accessed 22 November 2011.
31. Muungano Support Trust (2011) The Mathare Zonal Plan. Available: http://muunganosupporttrust.wordpress.com/2011/10/28/the-mathare-zonal-plan/. Accessed 11 November 2011.
32. World Bank (2011) Kenya Informal Settlements Improvement Programme (KISIP). Available: http://web.worldbank.org/external/projects/main?Projectid=P113542&theSitePK=40941&piPK=73230&pagePK2=64283627&menuPK=228424. Accessed 3 December 2011.
33. City of Richmond (2011) Community Health and Wellness Element. Available: http://www.ci.richmond.ca.us/index.aspx?NID=2412. Accessed 25 April 2012.
34. Toronto Central Local Health Integration Network (LHIN) (2011) Toronto Central LHIN website. Available: http://www.torontocentrallhin.on.ca. Accessed 13 November 2011.

35. Centre for Health Promotion (CEDAPS) (2011) CEDAPS website. Available: http://www. cedaps.org.br. Accessed 12 October 2011.
36. Belo Horizonte Observatory for Urban Health (OSUBH) (2011) OSUBH website. Available: http://www.medicina.ufmg.br/osubh. Accessed 9 December 2011.
37. India Ministry of Health (2011) National Urban Health Mission. Available: http://mohfw. nic.in/NRHM/Documents/Urban_Health/UH_Framework_Final.pdf.   Accessed   24 November 2011.
38. San Francisco Department of Public Health (2011) The Healthy Development Measurement Tool. Available: http://www.thehdmt.org. Accessed 22 September 2011.

Supplemental tables and figures available at http://journals.plos.org/plosmedicine/article?id=10.1371/ journal.pmed.1001285.

PART II

Citizen Engagement in Land-Use Decisions

CHAPTER 4

Owning the City: New Media and Citizen Engagement in Urban Design

Michiel De Lange and Martijn De Waal

4.1   INTRODUCTION

In today's cities our everyday lives are shaped by digital media technologies such as smart cards, surveillance cameras, quasi–intelligent systems, smartphones, social media, location–based services, wireless networks, and so on. These technologies are inextricably bound up with the city's material form, social patterns, and mental experiences. As a consequence, the city has become a hybrid of the physical and the digital. This is perhaps most evident in the global north, although in emerging countries, like Indonesia and China mobile phones, wireless networks and CCTV cameras have also become a dominant feature of urban life (Castells, et al., 2004; Qiu, 2007, 2009; de Lange, 2010). What does this mean for urban life and culture? And what are the implications for urban design, a discipline that has hitherto largely been concerned with the city's built form?

In this contribution we do three things. First we take a closer look at the notion of 'smart cities' often invoked in policy and design discourses about the role of new media in the city. In this vision, the city is mainly understood as a series of infrastructures that must be managed as efficiently as possible. However, critics note that these technological imaginaries of a personalized, efficient and friction–free urbanism ignore some of the basic tenets of what it means to live in cities (Crang and Graham, 2007).

Second, we want to fertilize the debates and controversies about smart cities by forwarding the notion of 'ownership' as a lens to zoom in on what we believe is the key question largely ignored in smart city visions: how to engage and empower citizens to act on complex collective urban problems? As is explained in more detail below, we use 'ownership' not to refer to an exclusive proprietorship but to an inclusive form of engagement, responsibility and stewardship. At stake is the issue how digital technologies shape the ways in which people in cities manage coexistence with strangers who are different and who often have conflicting interests, and at the same time form new collectives or publics around shared issues of concern (see, for instance, Jacobs, 1992; Graham and Marvin, 2001; Latour, 2005). 'Ownership' teases out a number of shifts that take place in the urban public domain characterized by tensions between individuals and collectives, between differences and similarities, and between conflict and collaboration.

Third, we discuss a number of ways in which the rise of urban media technologies affects the city's built form. Much has been said and written about changing spatial patterns and social behaviors in the media city. Yet as the editors of this special issue note, less attention has been paid to the question how urban new media shape the built form. The notion of ownership allows us to figure the connection between technology and the city as more intricate than direct links of causality or correlation. Therefore, ownership in our view provides a starting point for urban design professionals and citizens to reconsider their own role in city making.

Questions about the role of digital media technologies in shaping the social fabric and built form of urban life are all the more urgent in the context of challenges posed by rapid urbanization, a worldwide financial crisis that hits particularly hard on the architectural sector, socio–cultural shifts in the relationship between professional and amateur, the status of expert knowledge, societies that face increasingly complex 'wicked' problems, and governments retreating from public services. When grounds are shifting, urban design professionals as well as citizens need to reconsider their own role in city making.

4.2   RECOUNTING THE ROLE OF URBAN TECH: FROM SMART CITY TO SOCIAL CITY

4.2.1   The personalized and efficient city

Urban media technologies stimulate a profound personalization of city life on spatial, social, and mental levels [1]. For example, on the spatial level GPS-enabled devices and navigation software enable quick familiarization with unknown terrain. On location-based platforms users check-in at particular locales, quickly grasp what is there and build up personal relationships with places (like becoming 'mayor'). Developments of what is known as the *Internet of Things*, or *Ambient Intelligence*, allow the automation of physical environments to respond to individual preferences [2]. On the social level, mobile communications allow people to continually keep in touch with their in-group (Licoppe, 2004; Ito, 2005), imagine a sense of nearness and intimacy [3], and solidify established relationships with friends and family at the expense of weak ties and strangers [4]. On the mental level, mobile devices with their multimedia capabilities allow people to create highly idiosyncratic images of the city [5]. Listening to music on one's mobile device for example generates—in the words of one of Michael Bull's respondents—the "illusion of omnipotence" [6]. These media thus foster an individualized 'sense of place', a feeling of being part and in control of a situation (Meyrowitz, 1985).

The push towards an efficient and personalized city is institutionalized on a much larger scale in smart city policies (Mitchell, 1999; Mitchell, 2006; Hollands, 2008; Allwinkle and Cruickshank, 2011; Ratti and Townsend, 2011; Chourabi, et al., 2012) [7]. Municipalities form alliances with technology companies and knowledge institutions with the aim to organize urban processes efficiently (for a recent research/policy agenda see Batty, et al., 2012). Sensor and network technologies gauge and optimize energy and water supplies, transport and logistics, air and environmental quality. The hope is that this improves the quality of life and that it helps to tackle some of the big future challenges that cities face. Companies that work on smart city strategies include IBM (http://www.ibm.com/thesmartercity), CISCO (http://www.cisco.com/web/strategy/smart_connected_communities.html), General Electric (http://www.gereports.com), AT&T (http://www.corp.att.com/stateandlocal/), Microsoft and Philips.

Examples of actual 'smart cities' include towns built from scratch like New Songdo in South Korea (http://www.songdo.com) and Masdar in the United Arab Emirates (http://masdarcity.ae), but more often existing cities that are made 'smarter', like the Amsterdam Smart City project in the Netherlands (http://amsterdamsmartcity.com).

4.2.2 Critique

As we note elsewhere (de Lange and de Waal, 2012a), the omnipresence of new media in an urban context has come under criticism along three broad lines. First, observers note that wayfinding devices, location-based services, digital signage, and customer loyalty cards transform our cities into consumer-optimized zones, while simultaneously producing exclusionary practices of 'social sorting' (Crang and Graham, 2007; Shepard, 2011; de Waal, 2012a, 2013). Second, omnipresent cameras with face and gait recognition software, RFID-based access cards, smart meters, connected databases, and mobile network positioning, push cities toward revived 'big brother' scenarios of pervasive institutional control and surveillance (Crang and Graham, 2007; Greenfield and Shepard, 2007; Lyon, 2009). Third, mobile screens, portable audio devices and untethered online access to one's familiar inner circle enable people to retreat from public life into privatized tele-cocoons, bubbles or capsules (Cauter, 2004; Habuchi, 2005; Bull, 2005; Ito, et al., 2009). In these scenarios city dwellers no longer engage with strangers around them. There is a lack of space for spontaneous encounters and public life, and a general lack of involvement with the immediate environment.

Additionally, 'smart city' developments take the technology lab as the starting point. The actual city is seen as the last and most difficult hurdle in successive phases of 'deployment' or 'roll-out', rather than the sole place where experiment truly proves its value. Smart city projects typically consist of a 'triple helix' of government, knowledge production (e.g., universities) and industry. Such consortia often ignore the role of citizens as equally important agents. At best citizens in smart city policies are allowed to provide feedback somewhere in the design process, although oftentimes they figure as 'end-users' instead of being engaged in the early stages of co-creation.

Artists and media activists have used these same media technologies to question and subvert the logic of the three Cs of consumption, control, and capsularization (de Lange and de Waal, 2012b) and approach urbanites as citizens rather than as consumers or end–users. This often happens through ludic interventions that hark back to Situationist legacies of dérive and detournement (Debord, 1958; Chang and Goodman, 2006; Charitos, et al., 2008; de Waal, 2012b). While we believe such criticisms are valuable, many remain highly temporary and stick to an oppositional politics. How can we use the potential strengths of urban technologies to help forge more durable 'project identities' [8]? We argue that an alternative take is needed on urban design with digital technologies that focuses on the active role of citizens and uses the city itself as the test bed for experiments.

4.2.3 'Social cities'

Another tale—still under construction—has recently risen to the fore. In this vision, urban technologies engage and empower people to become active in shaping their urban environment, to forge relationships with their city and other people, and to collaboratively address shared urban issues (Paulos, et al., 2008; Foth, et al., 2011; de Lange and de Waal, 2012b). The focus in these discussions is on 'social cities' rather than on 'smart cities' [9]. It explores how digital media technologies can enable people to act as co–creators of livable and lively cities. This narrative is inspired by the body of literature that describes profound shifts in the balance between production and consumption: from professional amateur to wisdom of the crowd, from do-it-yourself culture to the hacker ethic (Himanen, 2001; Leadbeater and Miller, 2004; Benkler and Nissenbaum, 2006; Shirky, 2008; Rheingold, 2012). Central is the question how collaborative principles and participatory ethics from online culture can be ported to the urban realm in order to coordinate collective action and help solve some of the urgent complex issues that cities are facing.

What then are these issues? These exist on multiple scales. Some have a global scope, like social equity and environmental sustainability, or adequate water, food and energy supplies. Others are specific to particular cities, like shrinking cities, aging populations and empty spaces. On an intermediary level many cities in the world face challenges such as the perceived decline of publicness, safety, social inclusion and cohesion, and the gap between citizens and policy. Such issues typically are not 'owned' by a single party. They are collective issues that involve multiple stakeholders and require forms of collaborative governance to tackle them. Typical for these issues is that short and long term interests of different stakeholders diverge. As a result it is hard to establish a common definition of the problem itself, let alone find a solution everyone agrees on. Moreover, a single intervention may catalyze unforeseen events that alter the initial state. Because of this complexity such issues have been called 'wicked problems' (Rittel and Webber, 1973).

4.3   OWNERSHIP: ENGAGING CITIZENS WITH NEW MEDIA

We want to contribute to the social city discourse by advancing the notion of 'ownership' as a lens to look at how cities are made and remade with the help of digital media. 'Ownership' acts a heuristic device to make sense of the variety of developments that can be grouped under the social city label. We use ownership to refer to the degree to which city dwellers feel a sense of responsibility for shared issues and are taking action on these matters. As such it is a 'hack' of ownership in everyday parlance as being the proprietor of something, which gives the possessor the right to exclude someone else. When understanding ownership in more inclusive terms it means that one has the right to act upon an issue. It is this sense of ownership that we are after: not a contractual, proprietary ownership, but a sense of belonging to a collective place, commitment to a collective issue, and willingness to share a private resource with the collective in order to allow other citizens to act, without infringing on other people's right of ownership. In Lefebvre's terms this is the right to appropriation, which is clearly distinct from the right to property [10].

What is the advantage of looking at urban issues as ownership questions? It highlights how in cities there often is a discrepancy between formal juridical rights on individual or institutional levels and a collective sense of responsibility for the lived environment. As said, ownership can have an exclusive meaning as proprietorship ("mine not thine") with passively conferred rights. This is the case with purely private matters and purely public matters for which the state is the sole responsible body. Ownership can also have an inclusive meaning that involves stewardship of what belongs to all of us. It then demands a stance of collective engagement and action. This inclusive and active notion of ownership underlines that city life is not just a matter of avoiding friction but also requires the willingness to affect, that is to touch upon things and other people and to set something or someone in motion (Thrift, 2004; de Lange, 2013).

Another advantage is that ownership offers a fresh take on existing models for citizen engagement. The idea of engaging citizens in shaping their living circumstances is of course not new. In many western countries it has been around since the 1970s. Among town planners, for example, 'place making' has been a popular concept, whereby local people have their say within a community–driven process (Beyea, et al., 2009). Policy-makers, housing corporations, politicians and knowledge institutes have also taken up the subject of citizen engagement. We can identify two extremes: a top-down participation model and a bottom-up community model. Policy institutions use participation models to initiate projects in which citizens are invited to have a say, like in a town hall meeting. Some critics dismiss this as 'pseudo-participation' (Miessen, 2010), which is reminiscent of what Arnstein has called 'tokenism' (Arnstein, 1969). Politicians and government authorities give participation a nostalgic sugarcoating of inclusivity, democratic decision-making and solidarity. In doing so they are 'offloading' their own responsibilities (Institute for the Future, 2010). This is especially urgent in the context of the 'Big Society' policy concept devised by the U.K. Conservative party, which seeks to shift from big government to "a political system where people have more power and control over their lives." [11]

The community model attempts to foster a sense of togetherness that has roots in physical proximity or virtual presence of homogenous groups of people who share key aspects of their lives. It upholds ideals of neighboring, localness, small-scale, similarity and simplicity. However, Jane Jacobs among others pointed out that city dwellers typically reject small-town parochialism. Or as she outspokenly put it:

> Togetherness is a fittingly nauseating name for an old ideal in planning theory. This ideal is that if anything is shared among people, much should be shared. "Togetherness," apparently a spiritual resource of the new suburbs, works destructively in cities. The requirement that much shall be shared drives city people apart. [12]

In her view cities offer citizens the advantage to escape narrow social control of the small village, and obtain the freedom to choose their own lifestyles.

With the notion of ownership we position ourselves in response to earlier investigations of using ICTs for urban issues in what has been called 'community informatics' [13]. While we continue in the line of thought that ICTs can be used to help solve shared issues, we disagree on the centrality of the notion of community. Shin and Shin for example note that the notion of community is morally charged and problematic, yet argue for community as an ideal to keep striving for: "[P]ursuing community is not merely an idealistic, utopian project; rather, it is a realistic requirement for life." [14]. Community, we believe, need not be the sole or even necessary precondition to act on collective issues. In our view community is too reminiscent of small–scale and local ways of life instead of contemporary urban life. Instead we prefer the use of 'networked publics' (Varnelis, 2008), groups of people who convene around a shared 'matter of concern' in entities that may be more fleeting, composed of differences rather than being based on sameness, and organized in distributed networks rather than in 'natural' social bonds of locality, class, ethnicity, cultural identity, and so on [15].

Importantly, complex urban issues often transcend purely local interests. Tenacious urban issues involve a complex of stakeholders, composed of citizens themselves, but also authorities and policy-makers on multiple levels, housing corporations, a wide array of social organizations and knowledge institutes involved in urban affairs, as well as local and global businesses. Ownership provides a horizon for action in which each stakeholder reciprocally contributes to the whole on a different but equal base.

Thus, with ownership we seek to overcome the parochialism inherent in bottom–up community models and the paternalism of top–down institutional participation policies. How can new media enable a more participatory kind of city making, without falling in the trap of either participation models in which nothing essentially changes, or the anti–urban ideals of localism and "small-is-beautiful" implied by community models? The advent of digital media technologies in the urban sphere offers opportunities to organize citizen engagement neither in local bottom–up nor institutionalized top-down fashion, but in networked peer-to-peer ways. Instead of seeking consensus these tools allow room for managing differences. We have seen how urban new media are often perceived to alleviate and eliminate moments of uncertainty and tension inherent to urban life. It is easy to understand how that threatens what according to prominent urban theorists is the city's fragile quintessence, namely living among strangers and dealing with differences and serendipitous situations (Simmel, 1997; Wirth, 1938; Jacobs, 1992; Milgram, 1970; Sennett, 1976). We should note however that there is nothing inherently new (or wrong per se) with personalizing and smoothing out the city. Since the rise of the early modern metropolis urbanites in one way or another have tailored the city to their individual preferences. People orient to familiar physical elements to feel more secure (Lynch, 1960). They play intricate social avoidance games of disengagement, distraction and deceit (Goffman, 1959; Lofland, 1973). They adopt blasé attitudes as a way to cope with sensory overload (Simmel, 1997; Milgram, 1970). The challenge therefore in our view is to balance these stories of personalization and efficiency on the one hand and of building collectives based on differences and mutualism on the other hand.

Individuals must not only devise avoidance strategies but also cooperate in order to address the more complex issues that are part of city life.

4.4 PROMISING DEVELOPMENTS FOR STRENGTHENING CITIZEN OWNERSHIP

As mentioned, 'ownership' is related to social policies that have been around since the 1970s. Nonetheless we argue that new media afford several promising qualitative shifts with regard to the way people engage, empower, and act, and in addition how they manage shared issues and resources. First, on the level of resources and issues 'big data' and urban media allow for collective issues to be named and made visible in new ways. Second, on the level of engagement media art projects contribute to a 'sense of place', allowing people to see themselves as part of the urban fabric. Third, media technologies empower new 'networked publics': groups of people who organize themselves around collective issues. Fourth, in what can be called 'DIY urbanism', media technologies allow citizens to act in new ways, for instance design their own city and collectively govern urban affairs.

4.4.1 Resources and issues: The rise of a data commons

A current development is considering the city as an information-generating system. A variety of technologies collect an enormous amount and range of data. Consciously or unconsciously, citizens contribute to the accumulation of data through their uses of all kinds of products and services. As these data are being aggregated, they may become a 'data commons': a new resource containing valuable information for urban designers. Datasets can be used to bring out, visualize and manage collective issues. Preconditions for the establishment of a data commons include the availability of and access to open data, and the skills citizens have to use the data in a meaningful way. With the notion of ownership in mind one issue at stake is who has possession rights over these data. Are these a limited number of players (mostly governmental authorities and private companies) or can citizens too have access to these data in order to create interesting new applications and services. Examples include a number of app contests that have been organized by various municipalities in the Netherlands based on open data sets [16]. Not only is it possible to use aggregated data about urban practices to visualize collective issues, it is also possible to bring out individual contributions and usage of resources.

4.4.2   Engagement: Sense of place

To engage people with communally shared issues, it is essential that people envision themselves as part of the urban fabric, and understand that their individual actions make a difference to the common good. They also need to trust other urbanites to act accordingly. Digital media can play an important part in this, and engage citizens in new ways. Various experiments have been done with this. Art projects like *Urban Tapestries* (http://urbantapestries.net) or the *Dutch Het geheugen van Oost* (*The Memory of Amsterdam East*, http://www.geheugenva-noost.nl) collect stories from various citizens and function as an exchange platform for these. Other projects such as Christian Nold's *Biomapping* (http://biomapping.net) act as provocative conversation pieces. Nold's installation collected biometric data from citizens while walking across town. The results—sudden spikes in heart rate or galvanic skin response—were used to engage locals in discussions about these places and the sensations they produced in them. Placeblogs have started to play a role in mapping diverse local initiatives in a particular area and by doing so produce a site where some of the stories of different people may start to overlap (Lindgren, 2005).

4.4.3   Publics: Networked publics

'Networked publics' are groups of people that use social media and other digital technologies to organize themselves around collective goals or issues (Varnelis, 2008). In online culture, networks of 'professional amateurs' create 'user generated content' or take part in 'citizen science' projects. Think of open source software or Wikipedia as successful examples. In cities we have seen a growing interest in organizing publics in such a way, either to collectively map issues as part of activism or to organize themselves around common pool resources. The Dutch Geluidsnet (http://geluidsnet.nl/en/) is an example of the former, in which citizens who live near Schiphol airport in the Netherlands started a campaign against excessive airport noise pollution. Participants set up a mesh network by installing sound sensors in or around their houses. This data was collected and aggregated to produce a body of facts that could be used as counter-evidence in their case against the airport. Lately we have seen a great interest in the organization of publics around so-called 'common pool resources' (Ostrom, 1990). These vary from car sharing and tool lending to urban gardening. What is new is that digital media make it easier to register individual contributions and usage of collective resources, and the reputation systems that emerge from these patterns may prevent the proverbial 'tragedy of the commons' (Hardin, 1968). What both these new interfaces have in common is that they make it easier to take a collective ownership into an issue or a common resource.

4.4.4   Act: DIY urban design

Digital media have enabled mechanisms for managing collective action. Traditionally, collectives suffer from a lack of information leading to less than optimal decision–making, which hampers action. With mobile and location-based media people can share more information more quickly and base adaptive decisions on it. Examples are the real–time exchange of information about air quality using portable sensors and mobile networks, or aggregated location–based information that allows predicting and providing information about traffic congestion. The terms 'co-creation' and 'crowdsourcing' are used for collective issues being tackled and managed collaboratively, with new participants having an active role. An interesting project is *Face Your World* (http://www.faceyour-world.net) by artist Jeanne van Heeswijk and architect Dennis Kaspori. Young people and other people living in an Amsterdam neighborhood collaborated in designing a city park using a 3D simulation environment in which they could upload their own images and ideas to debate amongst each other. With this crowdsourced plan they managed to persuade the local government to abandon the initial plans for the park and execute theirs instead. Like online counterparts that successfully manage collective action (from Wikipedia to the Linux kernel), it would be an illusion to view these phenomena as exclusively bottom-up pro-cesses. They require curatorship and sets of rules. These rules are oftentimes enforced not by singular top-down institutions but through distributed forms of supervision and sanctions organized by users themselves.

4.4.5   Limitations of 'ownership'

The lens of ownership also brings out a number of problematic issues with regard to the social organization of urban life with the help of new media. Many of the examples above are still anecdotal. Others have their origin in the domain of art. Both show that urban media do have the affordance to promote 'owner-ship'. However, the examples provided also raise pertinent and interrelated questions: what is the effectiveness or social merit of these interventions, and how do we institutionalize these new forms? Once new urban issues have been visualized, and an initial interest or sense of engagement is aroused, how can publics organize in a productive way around them? What legal and regulatory frameworks do we need for instance to allow citizens to produce their own energy in a collaborative structure and deliver their surplus to the grid? What new types of institutions are needed and how can the pitfalls of utopian new society-making be avoided? By taking these questions as points of departure, 'ownership' can also be used as a design and policy approach that offers an alternative to the urban imaginary of 'smart cities'.

4.5 IMPLICATIONS FOR URBAN DESIGN: NEW MEDIA AND THE BUILT FORM

The relationship between (digital) media technologies and the physical city has often been thought of in a straightforward, even simplistic manner. The relation has long been theorized in terms of a substitution effect whereby ICTs eventually would make the physical urban form obsolete [17]. In this view, voiced by, for instance, McLuhan, Virilio, and Mitchell, ICTs would lead the city to become increasingly dematerialized, decentralized and ephemeral [18]. ICTs would cause the disappearance of concentrated functions from the city centers in realms such as commerce (Dodge, 2004), public institutions (Mitchell, 1995), and housing [19]. To be fair it should be added that de Sola Pool takes a more nuanced approach than depicting technology's impact on the city as merely one-way. Despite its title, de Sola Pool and his colleagues make it consistently clear in *The social impact of the telephone* (1977) that the telephone is "a facilitating device" and that it "often contributed to quite opposite developments" [20]. The city and the telephone 'mutually shape' or modify each other. The telephone (and the car) "were jointly responsible for the vast growth of American suburbia and exurbia, and for the phenomenon of urban sprawl. There is some truth to that, even though everything we have said so far seems to point to the reverse proposition that the telephone made possible the skyscraper and increased the congestion downtown" [21]. Since the early 1990s onwards a growing number of authors have pointed out that ICTs actually concentrate functions and people in cities. Cities are hubs for information networks, skills and knowledge in 'global cities' and 'technopoles' (Sassen, 1991; Castells and Hall, 1994) and for cultural industries in 'creative cities' (Florida, 2004).

At the level of design practice crude translations from observation to intervention frequently result in slavishly catering to some of the technological affordances discussed in the first section. For instance in reaction to people working ubiquitously with their portable wireless devices, a host of spaces are adapted to nomadic labor by being equipped with Wi-Fi, power sockets and cocooning zones. Convenient as this may be for individuals, such a reactive, even servile attitude of urban design to the demands of 'technological progress' avoids a more critical engagement that interrogates the desirability of such developments (de Lange and de Waal, 2009).

We believe it is necessary to explore alternatives to direct connections of causality or correlation between technology and the city. Ownership allows us to venture beyond relationships of amplification, substitution, or modification, and take a more culturally sensitive detour that highlights new ways of co-creating the city.

For one, the data generated by the city can be used as variables in (parametric) design approaches. Architects and other professionals can and are already using these data to gain insight in spatial patterns of citizens, about their mental maps and emotional sense of well–being tied to particular places, or to learn about the presence or absence of particular subcultures to whom designs can be tailored. Dutch architecture and research office Space&Matter (http://www.spaceandmatter.nl/index.php/architecture/urban-eindhoven/) harvested social network data to research a transformation plan for an old energy plant in Eindhoven. Through these searches they found two subcultures of skaters and BMX bikers, and climbers. By investigating and comparing their respective spatial needs, they proposed to strike a balance in the reuse of the building by retrofitting it with perforations in the floor that would benefit both subcultures.

The data that the city and its inhabitants produce can be used to visualize collective issues in new ways that appeal to people's emotional attachment. For instance, there have been quite a few projects trying to visualize environmental issues, from MIT's Senseable City Lab's Trash Track (http://senseable.mit.edu/trashtrack/), which follows the route of discarded objects, to the Medialab Prado's *In the Air* (http://www.intheair.es/), which measures and displays air pollution. Most data visualization projects stay in the digital realm of 'information architecture', turning data in beautiful visualizations. Some of them however jump over to urban architecture by experimenting with physical and tangible installations rather than online maps or projections on museum walls. For *In the Air* a prototype was developed for a fountain with colors and light intensity that reflect air quality. In the Dutch city of Doetichem artist Q.S. Serafijn and architect Lars Spuybroek created the D-Tower (http://www.d-toren.nl/site/), an interactive light sculpture that reflects the mood of the city and which can be seen as an early exploration of an 'architecture of affect' (see de Lange, 2013). The colors of the light installation (yellow for fear, green for hatred, red for love and blue for happiness) are determined by the outcomes of a daily online questionnaire amongst residents about their mood. As the project was finalized in 2004 it did not yet make use of any real–time information. It can be expected that in the near future many interactive installations, light sculptures and other objects will appear in the city that reflect in concrete or more abstract ways the real-time rhythms and emotions of the city or address particular issues (such as air pollution) that may arise from the data commons.

At the same time we witness the emergence of new spatio-temporal types. For some time now many cities have seen so-called "pop-up" events (pop-up bars, pop-up clubs, pop-up shops), often in vacant buildings and underused sites (Schwarz and Rugare, 2009). Additionally, crowdfunded neighborhood buildings and infrastructures emerge that are sometimes literally built with second hand or discarded materials (an example in Amsterdam is http://noorderparkbar.nl). Often organized with a collaborative DIY attitude and with the aid of social media, these interventions shift focus from place making to creating temporary events. Their sudden appearance and impermanence underline the transient nature of urban places in an age of new media developments that occur on a completely different timescale from traditional architecture (de Lange and de Waal, 2009). Thus, the balance of architectural practice appears to shift from manipulating space to manipulating space in time. A case taken to the extreme is DUS Architect's Bubble Building (http://dusarchitects.com/projects.php?categorieid=publicbuildings&projectid=bubblebuilding) made entirely out of soap bubbles. It is meant to stimulate playful interactions since visitors must collaborate to build the soap structure.

In these examples we see how some of the tensions mentioned in the introduction—individual and collective, difference and similarity, conflict and collaboration—become materialized and reconfigured in architecture. The rise of urban data means it is much easier to find, build and live among people based on perceived similarities. This is partly true in the case of collective private commissioning (CPC), an official Dutch housing policy measure since 2000 that aims to stimulate end-users to collectively design and build their own homes, as they had prior to World War II after which public housing became the task of national government, local authorities and semi-public housing corporations.

CPC aims to fit the mobility and DIY attitude of the present network society, and "the need for a renewed collective self-esteem" [22]. While on the scale of the housing project this may lead to homogenization, as likeminded people tend to cluster and choose similar designs, it may lead to a mosaic-like heterogeneity at the wider scale of neighborhoods. Nonetheless it raises questions about who owns the city, as an evaluative study into ten years of CPC and variants finds: "[A]ccording to the residents questioned, there are some cases where (C)PC projects seem to be perceived as 'different' and 'gated'. Although openness is often guaranteed, some are still regarded as outsiders." [23].

In the above cases traditional institutions are often bypassed. Architects adopt the roles of commissioner and executor at once. Rather than being demand-driven and waiting for a commission or entering competitions, they actively seek out an issue like the redevelopment or temporary use of a particular place and try to organize publics that take ownership. Instead of pitching they campaign and mobilize networked publics to realize their plans. This movement away from a demand-driven work ethic appears to have striking parallels with the intrinsically motivated playful hacker spirit of doing something just because it is fun [24].

4.6   CONCLUSION

We have forwarded 'ownership' as a lens to look at the role of new media technologies in the city, chiefly as an alternative to the smart city paradigm. We have shown how digital media have created a number of qualitative shifts in the way publics can be engaged with, organized around and act upon collective issues. These shifts mean that it has become easier for many citizens to organize themselves and take ownership of particular issues. In turn this may lead not only to new ways in which social life is organized, but also to new ways of shaping the built environment. We also argued that a culturally sensitive approach to the relation between city and technology is much needed. While many of these developments spring from grassroots initiatives and are organized around decentralized networks, they certainly are not without structure, rules and institutions. Of course we have to keep in mind that not everyone has access to these digital technologies, let alone is 'net smart' enough to use them beneficially (Rheingold, 2012). Another issue for further debate is the ongoing struggle over control of infrastructures and data. Perhaps this is a contribution architects and other urban designers can make to the world of new media design: to design truly accessible and inclusive urban interfaces that engage citizens with particular issues and allow to them to organize themselves and act.

FOOTNOTES

1. Ling, 2008; Paulos, et al., 2008; de Lange, 2010: pp. 179–183; Dourish and Bell, 2011; de Waal, 2012a.
2. In the words of a company that sells Near Field Communication solutions, this will produce an "effective personalization of the physical world". Source: http://www.nearfieldcommunication.com/business/overview/, accessed 23 September 2012.
3. de Gournay, 2002: pp. 201–204; Fox, 2006, p. 13.
4. Ling, 2008, pp. 159, 182.
5. Bull, 2005; de Lange, 2009, p. 66.
6. Bull, 2005, p. 175.
7. See also numerous special journal issues about smart cities, like Journal of Urban Technology (volume 18, number 2, 2011); Urbanist (number 517, 2012); Journal of the Knowledge Economy (volume 4, number 2, 2013); Economist (27 October 2012).
8. Manuel Castells distinguishes between the dominant 'legitimizing identity', the counter–active 'resistance identity', and the affirmative 'project identity' (Castells, 1997, pp. 7–8).
9. See the documentation on the international workshop and conference "Social Cities of Tomorrow", organized by The Mobile City, Virtueel Platform and ARCAM, 14–17 February 2012 in Amsterdam, www.socialcitiesoftomorrow.nl.
10. Lefebvre, 1996, p. 174; Mitchell, 2003, p. 18; Pugalis and Giddings, 2011, p. 282.
11. Conservative Party (Great Britain), 2010, p. ix.
12. Jacobs, 1992, p. 62.
13. Gurstein, 2000, 2003; Keeble and Loader, 2001; Foth, 2009: p. xxix; Shin and Shin, 2012.
14. Shin and Shin, 2012, p. 28.
15. See also Latour, 2005, p. 114.
16. See, for instance, Apps for Amsterdam (www.appsforamsterdam.nl/en).
17. For critical discussions, see Downey and McGuigan, 1999; Graham, 2004, pp. 3–24; Picon, 2008, pp. 32–34; de Lange, 2010, pp. 160–166; Tuters and Lange, 2013.
18. McLuhan, 1994, p. 366, pp. 378–379; Mitchell, 1995; Virilio, 1997, p. 25.
19. de Sola Pool, 1977, pp. 141, 302.
20. Pool, 1977, p. 302.
21. Pool, 1983, pp. 43–44.
22. Boelens and Visser, 2011, pp. 105–106.
23. Boelens and Visser, 2011, p. 124.
24. Himanen, 2001, pp. 3–7.

REFERENCES

1. Sam Allwinkle and Peter Cruickshank, 2011. "Creating smart–er cities: An overview," Journal of Urban Technology, volume 18, number 2, pp. 1–16. doi: http://dx.doi.org/10.1080/10630732.2011.601103, accessed 5 November 2013.
2. Sherry R. Arnstein, 1969. "A ladder of citizen participation," Journal of the American Planning Association, volume 35, number 4, pp. 216–224.

3. Michael Batty, Kay Axhausen, Fosca Giannotti, Alexei Pozdnoukhov, Armando Bazzani, Monica Wachowicz, Georgios Ouzounis, and Yuval Portugali. 2012. "Smart cities of the future,"European Physical Journal Special Topics, volume 214, number 1, pp. 481–518. doi: http://dx.doi.org/10.1140/epjst/e2012-01703-3, accessed 5 November 2013.

4. Yochai Benkler and Helen Nissenbaum. 2006. "Commons–based peer production and virtue," Journal of Political Philosophy volume 14, number 4, pp. 394–419. doi: http://dx.doi.org/10.1111/j.1467-9760.2006.00235.x, accessed 5 November 2013.

5. Wayne Beyea, Christine Geith, and Charles McKeown. 2009. "Place making through participatory planning," In: Marcus Foth (editor). Handbook of research on urban informatics: The practice and promise of the real–time city. Hershey, Pa.: Information Science Reference, pp. 55–57. doi: http://dx.doi.org/10.4018/978-1-60566-152-0.ch004, accessed 5 November 2013.

6. Luuk Boelens and Anne–Jo Visser. 2011. "Possible futures of self–construction: Poststructural reflections on ten years of experimentation with (C)PC," In: Lei Qu and Evert Hasselaar (editors). Making room for people: Choice, voice and liveability in residential places. Amsterdam: Techne Press, pp. 103–128.

7. Michael Bull, 2005. "The intimate sounds of urban experience: An auditory epistemology of everyday mobility," In: János Kristóf Nyíri (editor). A sense of place: The global and the local in mobile communication. Vienna: Passagen Verlag, pp. 169–178.

8. Manuel Castells. 1997. The power of identity. Malden, Mass.: Blackwell.

9. Manuel Castells and Peter Hall. 1994. Technopoles of the world: The making of twenty–first–century industrial complexes. London: Routledge.

10. Manuel Castells, Mireia Fernandez–Ardevol, Jack Linchuan Qiu, and Araba Sey. 2004. The mobile communication society: "A cross–cultural analysis of available evidence on the social uses of wireless communication technology," Annenberg Research Network on International Communication, University of Southern California, at http://arnic.info/workshop04/MCS.pdf, accessed 5 November 2013.

11. Lieven de Cauter, 2004. The capsular civilization: On the city in the age of fear. Rotterdam: NAi Publishers.

12. Michele Chang and Elizabeth Goodman, 2006. "Asphalt games: Enacting place through locative media," Leonardo, volume 14, number 3, at http://www.leoalmanac.org/wp-content/uploads/2012/07/Asphalt-Games- Enacting-Place-Through-Locative-Media-Vol-14-No-3-July-2006-Leonardo-Electronic-Almanac.pdf, accessed 5 November 2013.

13. Dimitris Charitos, Olga Paraskevopoulou, and Charalampos Rizopoulos. 2008. "Location-specific art practices that challenge the traditional conception of mapping," Artnodes, number 8, at http://www.uoc.edu/artnodes/8/dt/eng/paraskevopoulou_charitos_rizopoulos.html, accessed 5 November 2013.

14. Hafedh Chourabi, Taewoo Nam, Shawn Walker, J. Ramon Gil-Garcia, Sehl Mellouli, Karine Nahon, Theresa A. Pardo, and Hans Jochen Scholl, 2012. "Understanding smart cities: An integrative framework," HICSS '12: Proceedings of the 2012 45th Hawaii International Conference on System Sciences, pp. 2,289–2,297. doi: http://dx.doi.org/10.1109/HICSS.2012.615, accessed 5 November 2013.

15. Conservative Party (Great Britain), 2010. "Invitation to join the government of Britain: The Conservative Manifesto 2010," [London: Conservative Party], at http://media.conservatives.s3.amazonaws.com/manifesto/cpmanifesto2010_lowres.pdf, accessed 5 November 2013.

16. Mike Crang and Stephen Graham, 2007. "Sentient cities: Ambient intelligence and the politics of urban space," Information, Communication & Society, volume 10, number 6, pp. 789–817. doi: http://dx.doi.org/10.1080/13691180701750991, accessed 5 November 2013.

17. Chantal de Gournay, 2002. "Pretense of intimacy in France," In: James E. Katz and Mark A. Aakhus (editors). Perpetual contact: Mobile communication, private talk, public performance. Cambridge: Cambridge University Press, pp. 193–205.

18. Guy Debord, 1958. "Theory of the Dérive," Internationale Situationiste, number 2, at http://www.cddc.vt.edu/sionline/si/theory.html, accessed 5 November 2013.

19. Martin Dodge, 2004. "Geographies of e–commerce: The case of Amazon.com," In: Stephen Graham (editor). The cybercities reader. London: Routledge, pp. 221–225.

20. Paul Dourish and Genevieve Bell. 2011. Divining a digital future: Mess and mythology in ubiquitous computing. Cambridge, Mass.: MIT Press.

21. John Downey and Jim McGuigan (editors), 1999. Technocities. London: SAGE.

22. Richard L. Florida, 2004. The rise of the creative class: And how it's transforming work, leisure, community and everyday life. New York: Basic Books.

23. Marcus Foth (editor), 2009. Handbook of research on urban informatics: The practice and promise of the real–time city. Hershey, Pa.: Information Science Reference.

24. Marcus Foth, Laura Forlano, Christine Satchell, and Martin Gibbs (editors), 2011. From social butterfly to engaged citizen: Urban informatics, social media, ubiquitous computing, and mobile technology to support citizen engagement. Cambridge, Mass.: MIT Press.

25. Kate Fox, 2006. "Society: The new garden fence," In: The mobile life report 2006: How mobile phones change the way we live, at http://www.yougov.co.uk/extranets/ygarchives/content/pdf/CPW060101004_1.pdf, accessed 5 November 2013.

26. Erving Goffman, 1959. The presentation of self in everyday life. Garden City, N.Y.,: Doubleday.

27. Stephen Graham (editor), 2004. The cybercities reader. London: Routledge.

28. Stephen Graham and Simon Marvin, 2001. Splintering urbanism: Networked infrastructures, technological mobilities and the urban condition. London: Routledge.

29. Adam Greenfield and Mark Shepard. 2007. "Urban computing and its discontents," Situated Technologies Pamphlet Series, at http://www.situatedtechnologies.net/files/ST1-Urban_Computing.pdf, accessed 12 December 2007.

30. Michael Gurstein (editor), 2000. Community informatics: Enabling communities with information and communications technologies. Hershey, Pa.: Idea Group Pub.

31. Michael Gurstein, 2003. "Effective use: A community informatics strategy beyond the Digital Divide," First Monday, volume 8, number 12, at http://firstmonday.org/article/view/1107/1027, accessed 5 November 2013.

32. Ichiyo Habuchi, 2005. "Accelerating reflexivity," In: Mizuko Ito, Misa Matsuda, and Daisuke Okabe (editors). Personal, portable, pedestrian: Mobile phones in Japanese life. Cambridge, Mass.: MIT Press, pp. 165–182.

33. Garrett Hardin, 1968. "The tragedy of the commons," Science, volume 162, number 3859 (13 December), pp. 1,243–1,248, and at http://www.sciencemag.org/content/162/3859/1243.full, accessed 5 November 2013. doi: http://dx.doi.org/10.1126/science.162.3859.1243, accessed 5 November 2013.

34. Pekka Himanen, 2001. The hacker ethic, and the spirit of the information age. New York: Random House.

35. Robert G. Hollands, 2008. "Will the real smart city please stand up? Intelligent, progressive or entrepreneurial?" City, volume 12, number 3, pp. 303–320. doi: http://dx.doi.org/10.1080/13604810802479126, accessed 5 November 2013.

36. Institute for the Future, 2010. "A planet of civic laboratories: The future of cities, information, and inclusion," at http://iftf.me/public/SR-1352_Rockefeller_Map_reader.pdf, accessed 5 November 2013.

37. Mizuko Ito, 2005. "Intimate visual co–presence," 2005 Ubiquitous Computing Conference (Tokyo), at http://www.itofisher.com/mito/archives/ito.ubicomp05.pdf, accessed 5 November 2013.

38. Mizuko Ito, Daisuke Okabe, and Ken Anderson. 2009. "Portable objects in three global cities: The personalization of urban places," In: Rich Ling and Scott W. Campbell (editors). The reconstruction of space and time: Mobile communication practices. New Brunswick, N.J.: Transaction Publishers, pp. 67–87.

39. Jane Jacobs, 1992. The death and life of great American cities. New York: Vintage Books.

40. Leigh Keeble and Brian Loader (editors), 2001. Community informatics: Shaping computer–mediated social relations. London: Routledge.

41. Michiel de Lange, 2013. "The smart city you love to hate: Exploring the role of affect in hybrid urbanism," In: Dimitris Charitos, Iouliani Theona, Daphne Dragona, and Haris Rizopoulos (editors). The hybrid city II: Subtle rEvolutions, at http://www.bijt.org/wordpress/wp-content/uploads/2013/06/130524_HC2-Athens.pdf, accessed 5 November 2013.

42. Michiel de Lange, 2010. Moving circles: Mobile media and playful identities. Faculty of Philosophy, Erasmus University Rotterdam, Rotterdam, at http://www.cs.vu.nl/~eliens/download/read/moving-circles.pdf, accessed 5 November 2013.

43. Michiel de Lange, 2009. "From always on to always there: Locative media as playful technologies," In: Adriana de Souza e Silva and Daniel M. Sutko (editors). Digital cityscapes: Merging digital and urban playspaces. New York: Peter Lang, pp. 55–70, and at http://www.bijt.org/wordpress/wp-content/uploads/2009/12/fromalwaysontoalwaysthere_def02.pdf, accessed 5 November 2013.

44. Michiel de Lange and Martijn de Waal. 2012a. "Ownership in the hybrid city", at http://virtueelplatform.nl/english/news/ownership-in-the-hybrid-city/, accessed 5 November 2013.

45. Michiel de Lange and Martijn de Waal. 2012b. "Social cities of tomorrow: Conference text," at http://www.socialcitiesoftomorrow.nl/background, accessed 5 November 2013.

46. Michiel de Lange and Martijn de Waal, 2009. "How can architects relate to digital media? The Mobile City keynote at the 'Day of the Young Architect'," at http://www.themobilecity.nl/2009/12/06/how-can-architects-relate-to-digital-media-tmc-keynote-at-the-%E2%80%98day-of-the-young-architect%E2%80%99//, accessed 5 November 2013.

47. Bruno Latour, 2005. Reassembling the social: An introduction to actor–network–theory. Oxford: Oxford University Press.

48. Charles Leadbeater and Paul Miller. 2004. The pro–am revolution: How enthusiasts are changing our economy and society. London: Demos.

49. Henri Lefebvre, 1996. Writings on cities. Selected, translated, and introduced by Eleonore Kofman and Elizabeth Lebas. Cambridge, Mass.: Blackwell.

50. Christian Licoppe, 2004. "'Connected' presence: The emergence of a new repertoire for managing social relationships in a changing communication technoscape," Environment and Planning D: Society and Space, volume 22, number 1, pp. 135–156. doi: http://dx.doi.org/10.1068/d323t, accessed 5 November 2013.

51. Tim Lindgren, 2005. "Blogging places: Locating pedagogy in the whereness of Weblogs," Kairos volume 10, number 1, at http://www.technorhetoric.net/10.1/binder2.html?coverweb/lindgren/, accessed 5 November 2013.

52. Rich Ling, 2008. New tech, new ties: How mobile communication is reshaping social cohesion. Cambridge, Mass.: MIT Press.

53. Lyn H. Lofland, 1973. A world of strangers: Order and action in urban public space. New York: Basic Books.

54. Kevin Lynch, 1960. The image of the city. Cambridge, Mass.: Technology Press.

55. David Lyon, 2009. Identifying citizens: ID cards as surveillance. Cambridge: Polity.

56. Marshall McLuhan, 1994. Understanding media: The extensions of man. Cambridge, Mass.: MIT Press.

57. Joshua Meyrowitz, 1985. No sense of place: The impact of electronic media on social behavior. New York: Oxford University Press.

58. Markus Miessen, 2010. The nightmare of participation. New York: Sternberg Press.

59. Stanley Milgram, 1970. "The experience of living in cities," Science, volume 167, number 3924 (13 March), pp. 1,461–1,468. doi: http://dx.doi.org/10.1126/science.167.3924.1461, accessed 5 November 2013.

60. Don Mitchell, 2003. The right to the city: Social justice and the fight for public space. New York: Guilford Press.

61. William J. Mitchell, 1999. E-topia: "Urban life, Jim — but not as we know it". Cambridge, Mass.: MIT Press.

62. William J. Mitchell, 1995. City of bits: Space, place, and the infobahn. Cambridge, Mass.: MIT Press.

63. William J. Mitchell, 2006. "Smart City 2020," Metropolis (April), http://www.metropolis-mag.com/April-2006/Smart-City-2020/, accessed 5 November 2013.

64. Elinor Ostrom, 1990. Governing the commons: The evolution of institutions for collective action. Cambridge: Cambridge University Press.

65. Eric Paulos, R.J. Honicky, and Ben Hooker, 2008. "Citizen science: Enabling participatory urbanism," In: Marcus Foth (editor), 2009. Handbook of research on urban informatics: The practice and promise of the real–time city. Hershey, Pa.: Information Science Reference. doi: http://dx.doi.org/10.4018/978-1-60566-152-0.ch028, accessed 5 November 2013.

66. Antoine Picon, 2008. "Toward a city of events: Digital media and urbanity," In: Stephen Ramos and Neyran Turan (editors). New Geographies volume 0, pp. 32–43.

67. Ithiel de Sola Pool, 1983. Forecasting the telephone: A retrospective technology assessment. Norwood, N.J.: ABLEX.

68. Ithiel de Sola Pool (editor), 1977. The social impact of the telephone. Cambridge, Mass.: MIT Press.

69. Lee Pugalis and Bob Giddings, 2011. "A renewed right to urban life: A twenty–first century engagement with Lefebvre's initial 'cry'," Architectural Theory Review, volume 16, number 3, pp. 278–295. doi: http://dx.doi.org/10.1080/13264826.2011.623785, accessed 5 November 2013.

70. Jack Linchuan Qiu, 2009. Working–class network society: Communication technology and the information have–less in urban China. Cambridge, Mass.: MIT Press.

71. Jack Linchuan Qiu, 2007. "The wireless leash: Mobile messaging service as a means of control," International Journal of Communication, volume 1, number 1, at http://ijoc.org/index.php/ijoc/article/view/15, accessed 5 November 2013.

72. Carlo Ratti and Anthony Townsend. 2011. "The social nexus," Scientific American, volume 305 (September), pp. 42–48. doi: http://dx.doi.org/10.1038/scientificamerican0911-42, accessed 5 November 2013.

73. Howard Rheingold, 2012. Net smart: How to thrive online. Cambridge, Mass.: MIT Press.

74. Horst W.J. Rittel and Melvin M. Webber, 1973. "Dilemmas in a general theory of planning," Policy Sciences, volume 4, number 2, pp. 155–169. doi: http://dx.doi.org/10.1007/BF01405730, accessed 5 November 2013.

75. Saskia Sassen, 1991. The global city: New York, London, Tokyo. Princeton, N.J.: Princeton University Press.

76. Terry Schwarz and Steve Rugare (editors), 2009. Pop up city. Cleveland, Ohio: Cleveland Urban Design Collaborative, College of Architecture and Environmental Design, Kent State University.

77. Richard Sennett, 1976. The fall of public man. New York: W.W. Norton.

78. Mark Shepard (editor), 2011. Sentient city: Ubiquitous computing, architecture, and the future of urban space. Cambridge, Mass.: MIT Press.

79. Yongjun Shin and Dong–Hee Shin, 2012. "Community informatics and the new urbanism: Incorporating information and communication technologies into planning integrated urban communities," Journal of Urban Technology, volume 19, number 1, pp. 23–42. doi: http://dx.doi.org/10.1080/10630732.2012.626698, accessed 5 November 2013.

80. Clay Shirky, 2008. Here comes everybody: The power of organizing without organizations. New York: Penguin Press.

81. Georg Simmel, 1997. "The metropolis and mental life," In: David Frisby and Mike Featherstone (editors). Simmel on culture: Selected writings. Thousand Oaks, Calif.: Sage, pp. 174–186.

82. Nigel Thrift, 2004. "Intensities of feeling: Towards a spatial politics of affect," Geografiska Annaler: Series B, Human Geography, volume 86, number 1, pp. 57–78. doi: http://dx.doi.org/10.1111/j.0435-3684.2004.00154.x, accessed 5 November 2013.

83. Marc Tuters and Michiel de Lange. 2013. "Executable urbanisms: Messing with Ubicomp's singular future," In: Regine Buschauer and Katharine S. Willis (editors). Medialität und räumlichkeit — Multidisziplinäre perspektiven zur verortung der medien (Multidisciplinary perspectives on media and locality). Bielefeld: Transcript, pp. 49–70.

84. Kazys Varnelis (editor), 2008. Networked publics. Cambridge, Mass.: MIT Press.

85. Paul Virilio, 1997. Open sky. Translated by Julie Rose. London: Verso.

86. Martijn de Waal, 2013. The city as interface: How new media are changing the city. Rotterdam: NAI Uitgevers/Publishers Stichting.

87. Martijn de Waal, 2012a. "De stad als interface: Digitale media en stedelijke openbaarheid," Proefschrift Rijksuniversiteit Groningen.

88. Martijn de Waal, 2012b. "The ideas and ideals in urban media," In: Marcus Foth, Laura Forlano, Christine Satchell, and Martin Gibbs (editors). From social butterfly to engaged citizen: Urban informatics, social media, ubiquitous computing, and mobile technology to support citizen engagement. Cambridge, Mass.: MIT Press, pp. 5–20.

89. Louis Wirth, 1938. "Urbanism as a way of life," American Journal of Sociology, volume 44, number 1, pp. 1–24.

CHAPTER 5

Urban Ecological Stewardship: Understanding the Structure, Function and Network of Community-based Urban Land Management

Erika S. Svendsen and Lindsay K. Campbell

5.1 INTRODUCTION

Much of the literature on civic environmentalism focuses on national and global campaigns and actors. There is a great deal of analysis on how social movement organizations and international NGOs interact with nation-states, intergovernmental entities, and other transnational NGOs (Wapner 1995; Keck and Sikkink 1998; Dalton et al. 2003). While these relationships are both critical and relevant, it is no less important to explore the nature and nuances of locally based, urban environmental stewardship organizations. Comprised of both informal and formal organizations and networks, these groups interact at multiple scales ranging from the household, to neighborhood, to urban area, to cross-regional scales. Scholars are beginning to recognize the gap in our understanding about the structure, function, and relationship between these groups and to question whether theories based on national organizations are applicable at the sub-national scale. For example, a recent study of environmental organizations in North Carolina examined organizational networks, coalitions, issues focus, membership characteristics and participation, financial resources, organizational practices and formality, leadership, and media engagement (Andrews and Edwards 2005). In this paper, several similar issues are considered for urban ecology organizations comparing cities in the Northeastern United States.

Local is the primary scale where abstract environmental principals or values intersect immediate quality of life concerns. There is a vibrant "backyard" environmentalism in the United States that goes beyond NIMBYism and beyond the rubric of environmental justice to include groups that are proactively managing sections of the landscape and planning for sustainability, both in urban and rural areas (Grove and Burch 1997; Weber 2000; Dalton 2001; Agyeman and Evans 2003).

Yet, the literature on civic environmental organizational strategies tends to neglect stewardship as a role or strategy, focusing instead on lobbying, letter writing, media campaigns, protests, boycotts, sitins, and even internet-based tactics (Coban 2003). Urban land stewardship is a strategy that includes elements of direct action, self-help, and often education and community capacity building. Ideologically, it is less rooted in oppositional social movements and more in accessing the rights to space through collaborative, community-based resource management. A fair amount has been written about community-based resource management in rural areas and developing nations, but this paper hopes to highlight how the same principles are being pursued in urban areas in the U.S (Burch and Grove 1993; Westphal 1993).

Carmin et al. (2003) identified communication, leveraging, and community development as the three main strategies used by regional environmental NGOs. While stewardship, itself, clearly focuses on the latter of those three strategies, the support offered to stewardship groups by civil society intermediaries can include the other two strategies as well. This paper suggests that urban environmental stewardship combines land management with the desires of civil society, the private sector and government agencies. Dynamics between and across scales of action are important to consider in trying to understand and parse out the actors and relationships within the network of urban land stewardship (see Figure 1).

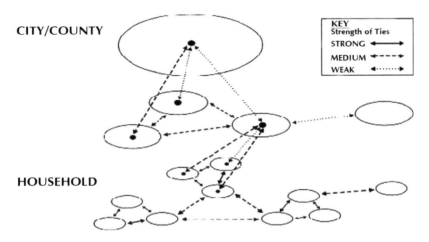

FIGURE 5.1   Multi-scaled model of Socio-Organizational Ties.
SOURCE: Grove et al. 2002.

In particular, this paper hopes to shed light on active organizations that are dedicated to using ecological strategies to create, restore, reveal or maintain any part of the urban landscape in six large urban areas in the Northeastern U.S.: Boston, New Haven, New York City, Pittsburgh, Baltimore, and Washington, D.C. These organizations include informal community groups, formal nonprofits, as well as municipal, state, and federal partners. While public, private and civil society entities will be discussed in this paper, each will be distinguished in the overall analysis. In order to support groups' stewardship efforts and improve their effectiveness as agents, a better understanding of their basic functioning as individual organizations and as a network is required. Using data from the Urban Ecology Collaborative (UEC) assessment, this paper examines how these organizations interact with critical biophysical resources (e.g. land, water, soil, air) and social institutions (e.g. government, commerce, education, non-profits) through the flow of materials, energy and information (e.g. human capital, funding, partnerships, science). The findings challenge three recent debates in urbanism, which claim that participation in civic associations is declining (Putnam 2000); that the urban environmental movement is place-based and fragmented (Harvey 1999) and that there is a waning public interest in issues pertaining to environmental quality (Greenberg 2005).

5.2   HUMAN ECOSYSTEM FRAMEWORK: A CONTEXT FOR UNDERSTANDING THE CHALLENGES OF URBAN STEWARDSHIP

Urban areas are ecosystems with interdependent resources and flows that are no less complex than wilderness or forested ecosystems (Burch and Grove 1993; Grove and Burch 1997; Pickett et al. 1997; Redman 1999). One might argue that in the urban context, the environment is nested within larger quality of life issues such as public health and well-being, economic development and social justice which are collectively driving social motivations for land based steward-ship. The Human Ecosystem Approach is used as a framework to aid in reveal-ing interactions that drive particular system at a particular point in time (see Figure 2). In this sense, the city-as-ecosystem is more than just a clever metaphor. Rather, it allows us to have a holistic understanding of the relationships between individuals, groups, organizations, culture, and norms—not just as sociological concerns, but as key contributors to the biophysical functioning of our cities. While one could choose any number of aspects from this Human Ecosystem

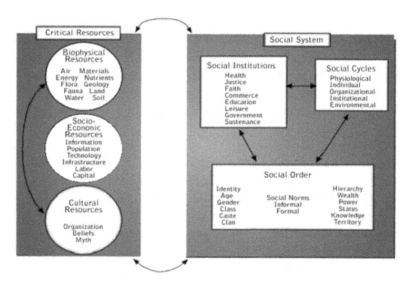

FIGURE 5.2   Human Ecosystem Framework.

SOURCE: Machlis, Force, and Burch 1997.

Framework (HEF) for study, this paper considers the role of organizations as a critical "cultural resource, for they provide the structural flexibility needed to create and sustain human social systems" (Machlis et al. 1997). Stewardship groups, in particular, are chosen because they are literally agents that interact with both the biophysical resources and the social system of the human ecosystem (see Figure 2). From practical or managerial standpoint, determining how best to manage the urban ecosystem requires a consideration of these human organizations as vital parts of the urban ecosystem.

Application of the HEF model to the analysis of the UEC assessment's stewardship organizations frames a number of key challenges that must be considered, which are described below.

5.2.1 Biophysical challenges

The largest percentage of the world's population living in urban areas was recorded at the turn of the 20th Century. It is no longer a question of whether urbanization affects ecosystem functions but rather, to what degree they do so and how positive externalities can be created within this highly manipulated system (Millennium Assessment 2005). At the metropolitan level, urban growth affects the heterogeneity of the landscape through landcover change and affects the spread of disturbance through invasive species, to name a few critical and documented examples (Alberti 2005). Within cities themselves, there is a range of open space areas from protected wildlife habitats, to contaminated and fallow sites, to highly managed and used parks. Habitats are fragmented, both discontinuous and small in size, yet species diversity can still be quite high in these disturbed landscapes (Niemela 1999). Basic urban infrastructure has major impacts on the environment. Landscape and social ecologists are still on the frontiers of knowledge regarding the management needs of highly urbanized areas. Yet, management and use of the landscape both by public authorities and the private sector continues regardless, despite a lack of understanding of how "best" to support certain ecosystem services. Often urban sites are not managed for biophysical function at all, instead serving social functions as recreation sites and as promoters of neighborhood efficacy (Sampson and Raudenbush 1999). It is in the interest of environmental planners to ascertain where, how and when community-based management of street trees, planter beds, lots, greenways, parks, and forests is occurring. And in urban areas, one simply cannot divorce the sites from their property jurisdiction, regulations, or users.

5.2.2 Social and organizational challenges

Stepping back from concerns about the ecosystem, there is a need for discourses on social capital, resource management and civic environmentalism to engage with the issue of urban stewardship, for it lies at a nexus of these issues. The debate over the proclaimed "death of associations" and accompanying dearth of social capital in American cities cites low membership in traditional civic and social groups like the American Legion, PTA and sports leagues (Skocpol and Fiorina 1999; Putnam 2000). In terms of the HEF model, Putnam is arguing that there is a decline in the (socioeconomic) resource of social capital as a function of where our society is in the current macro social cycles of participation and volunteerism, as influenced by media and technology like the television. While Putnam's hypotheses and methodology has been challenged, his contribution to the public perception of local involvement is great (Edwards and Foley 1998). To this critique, this paper adds another argument. A new class of ecologically-minded nonprofit and community based groups is emerging in urban areas as 69% of the civil society groups surveyed in the UEC assessment were formed in 1980 or later; and 55% of the civil society groups consider their service areas to be at the city or sub-city level. While Skocpol emphasizes the change rather than decline of civic environmental associations in the 1970s, the focus remains on groups organized nationally for direct political purpose (Skocpol and Fiorina 1999). These national organizations have been the basis of environmental organization research and typically have small constitutions at the local level. Both Putnam and Skocpol's work differs from the UEC research, which suggests that vital social organizations emerge and expand from local, place-based and laterally networked issues. At the same time, the UEC findings hint that environmental motivations are nested within larger quality of life issues.

Similarly, activists and scholars alike have proclaimed that we are experiencing "the death of environmentalism", citing the institutionalization of environmental non-profits, fragmentation and their inability to achieve necessary, radical environmental change (Harvey 1999; Shellenberger and Nordhaus 2004). A version of this argument read through the HEF model is: the environmental movement's current cultural resources are inadequate (or, misappropriated) to achieve its goals, given the existing social institutions (government, business) and the social order (power, hierarchy, norms). Authors focusing on national organizations and surveys are typically discussing issues at a particular scale, such as international climate change or environmental quality (Fisher 2004; Fisher and Green 2004; Greenberg 2005). Criticism therefore focuses on policy-oriented and broad membership organizations, which wholly ignores that the rhetoric of "death of environmentalism" is not relevant to community-based stewardship groups that are actively integrating biophysical and social goals. Evidence of this emerges in this assessment as groups straddle the divide between environmental protection and community development. Based on the coding of open-ended reporting of missions and major programs, these groups focus on improvement of environmental quality (22.5%), community development (39.2%), and environmental education (38.3%).

5.2.3   Collaboration challenges

Some of the most visible efforts at collaborative natural resource management occur in high profile land use conflicts in the Western United States. Many forest, rangeland, and coastal managers attempt to achieve stakeholder-inclusive, ecosystem scale management (Weber 2000; Wondolleck and Yaffee 2000; McCreary 2001). However, recent studies have shown that similar patterns of nonoppositional strategies are emerging within the urban frame (Sirianni and Friedland 2001). This suggests that while there may be a wide range of urban environmental actors using multiple strategies, there can be cohesive management and policy-making, given the time and space to negotiate. While partnership strategies and coalitions certainly exist, the concentrated problems—particularly in low income urban communities—of water quality, air quality, soil quality, availability and distribution of open space, and toxics far outpace the political power or organizational capacity of any single group to add them adequately (Bullard 1990). As such, there remains a great deal of work to be done in the coordination of urban ecosystem management. This is not to suggest that all management in cities should be centralized, but these findings suggest the need to recognize and harness the degree of diversity, autonomy and effectiveness among public and private sector stewardship regimes. Any attempt to understand who these groups are, why they are involved with caring for the urban landscape, and what can be done to help them work more effectively in light of the many challenges can increase the likelihood of coordinated urban ecosystem management.

5.3   METHODS

The assessment was conducted in 2004 by the research subcommittee of the UEC, with supporting funding from the USDA Forest Service. The goal of the assessment was to determine the status of organizations and community-based urban stewardship initiatives operating in selected major cities in the Northeastern U.S. Specifically, it intended to:

- "Discover the gaps between biophysical and social resources, organizations, and programs;
- Highlight specific stewardship opportunities, priorities and resources in each major city;
- Examine the current capacity of organizations to use urban and community forestry activities in the improvement of the physical environment and quality of life issues common to large urban areas;
- Determine strategies for the exchange of urban and community forestry tools and techniques." (UEC 2004)

There was some slight variation by city in terms of methodology; as the established process was that each city would generate (or use existing) lists of organizations that are currently engaged in urban ecology initiatives. These initiatives could range from tree planting, to open space design, to environmental education, with the common criterion being that the groups must be actively supporting or caring for a particular piece of the urban landscape. From these lists, a sample of organizations was selected for study, stratified by management type, which consisted of: non-profit, federal, state, and local government, for-profit, community-based groups and individuals (usually independent environmental contractors). The outreach strategy to those organizations varied by city: New Haven convened a meeting and distributed surveys in person; Pittsburgh, Washington, D.C., and Boston relied upon emailing and phone outreach. The New York City methodology is described here in greater detail, as it may be most useful as a model for expanding research on a more expansive sampling framework.

5.3.1 New York City Sampling Methodology

The sample of 100 organizations and informal groups for the New York City assessment was drawn from a population of 2,027 groups compiled from the combined stewardship databases, participant rosters, and organizations tracked by the largest urban ecology intermediary groups in the city and in some cases region. This chart represents the groups used for this assessment who were tracking explicit stewardship information.

- Partnerships for Parks: 1,000 active, park-based volunteer groups
- Council on the Environment for New York City (CENYC): 600 community gardens
- NYC Department of Parks and Recreation GreenThumb Program: 324 community gardens
- Harbor Estuary Program (HEP): 300 regional stewards

These core databases were supplemented with additional groups categorized as relating to environmental issues from the New York City Nonprofits Project citywide survey of projects, as well as attendees of meetings included in the Open Accessible Space Information System (OASIS), Metro Forest Council databases, listed partners from the Earthpledge website, and groups listed on the Neighborhood Open Space Coalition's Hub website.

After the databases were assembled, they were merged along all common characteristics and duplicate listings were eliminated. Then two fields "scale" (region; city; borough; neighborhood/block) and "management type" (public agency federal; public agency state; public agency city; for-profit; nonprofit; community group) were ascertained for each group, based on information in the existing databases and input from the staff of organizations maintaining the databases. Some unknowns remained for which management type and scale could not be determined, and these were excluded. The fields were then used to stratify the sample. A four percent sample was taken from all community groups and nonprofits. Because of the limited number of organizations, with many of the natural resource groups being known entities, federal, state, and local agencies were purposively over sampled in the assessment. Forprofit groups were randomly sampled. The sampling is summarized in Table 1.

The 100 selected groups were sent the survey by mail, with a follow-up phone call to answer any remaining questions, followed by a postcard reminder to complete the survey and one final round of calls, all conducted in the summer of 2004.[1] Of the surveyed organizations, 34 completed the survey, eight said the survey was not applicable to their group (because they were actually not engaged in stewardship), and one refused to participate. Clearly, community groups had the lowest response rate, which is not surprising given the challenge of reaching these informal and sometimes temporary groups. It is possible that a number of non-responses were due to groups that no longer exist, given the age of some of the stewardship databases comprising the parent population.

TABLE 5.1 Type of Environmental Management by Geographic Scale

MANAGEMENT TYPE

SCALE		Un-known	Public agency federal	Public agency state	Public agency local	For-profit	Non-profit	Comm group
	Unknown	1	0	0	0	0	42	3
	Region	1	16	8	0	7	87	1
	City	0	1	1	12	16	96	5
	Borough	1	0	2	7	0	74	25
	Neigh-borhood/ block	7	0	0	23	9	432	1150
	SUBTOTAL (excluding unknowns)		17	11	42	32	689	1181
	TOTAL= 2,027 including unknowns; 1,972 after excluding unknowns							
	Sampling Methodology		Purposive Selection	Purposive Selection	Purposive Selection (4) + Random Selection (3)	Random Selection	Random selection (4% of total), stratified by scale	Random selection (4% of total), stratified by scale
	Surveyed (n=100)		9	6	7	4	27	47
	Returned (n=34)		2	4	7	0	12	6 (+2 indv)

The six cities, combined to survey 135 organizations (34 in New York City, 19 in Baltimore, nine in Boston, 34 in Washington, D.C., 20 in New Haven, and 19 in Pittsburgh), is not comprehensive enough to make any sort of quantitative cross-city comparisons. Because the sample was not drawn randomly, it does not enable the use of predictive statistics (e.g. regressions or means testing) on this dataset. Although this limits the analysis and makes clear the need for further study, the intent of this project was to characterize the basic form and function of an under-studied set of civil society and public actors. Thus, frequencies and percentages will be used to report the overall trends in the data.

5.4 FINDINGS AND DISCUSSION

5.4.1 *Organizational Demographics: Management Type and Age of Organization*

Organizational demographics are some of the fundamental attributes of these groups, including management type and age. Because the goal of the UEC assessment was to understand local environmental stewardship, rather than solely the role of civil society, we see that there is a mix of organization types included in the results (see Figure 3). However, despite an attempt to be inclusive of government actors, it is evident that civil society actors outnumber them, with nonprofits, community groups, and individuals comprising 73% of the sample. This is likely a reflection of the fact that government agencies are larger and more centralized, while nonprofits and community groups are more local and place-based. So, for example, while there is one New York City Park Department, there are over 600 community gardens and more than 1000 active park-based stewardship groups in New York City.[2] The level of civil society involvement is significant from a managerial standpoint, since it means that resource managers wishing to make changes on a landscape or to improve ecological functioning in a watershed will need to do so in concert with informal and nonprofit groups. However, this does not suggest the absence of public sector involvement as suggested in the case of advocating for citizen monitoring "bucket brigades" (O'Rourke and Gregg 2003). Instead, it may suggest the need to reconsider models for shared stewardship or 'governance' of urban land (Durant 2004).

In fact, one could perhaps make the argument that the hard boundaries of public entities and civil society actors begin to blur at the local level. There are numerous examples of intermediaries: Partnerships for Parks is a public-private entity that is a combination of the New York City Parks Department and the City Parks Foundation, dedicated to supporting community groups in their engagement with parks; GreenThumb is a federal Community Development Block Grant (CDBG) supported program of the New York City Parks Department that offers resources, materials, and technical assistance directly to informal community gardening groups; and the Harbor Estuary Program is a National Estuary Program authorized by the EPA that includes participants "from local, state, and federal environmental agencies, scientists, citizens, business interests, environmentalists, and others" (Program 2002). These intermediaries, organized around particular site types, seem to have a more prominent presence in

New York City than the other cities studied, which is a function of the size and complexity of the stewardship network. These organizations differ from the majority of the small nonprofits and groups included in this survey that directly carry out volunteer stewardship. Their primary function is to maintain flows of material, information, and resources. They bear some resemblance to intermediaries that work in other areas of the urban environment, such as large CDCs or nonprofit coalitions that coordinate citywide brownfield inventories, such as the Cleveland Neighborhood Development Coalition (Brachman 2003).

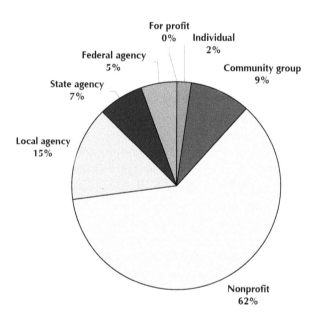

FIGURE 5.3   Type of Environmental Management.

Distinctly missing from this assessment is the business community. This is due to both to the nature of the populations from which the samples were drawn and the criterion applied for inclusion in the survey. The New York City parent population (the combined databases of the environmental intermediaries) illustrates the first issue, with just 32 for-profit entities out of the total 2,004 organizations, the business sector is simply not in this stewardship network as we sampled.[3] Second, the baseline criterion applied was that each respondent had to be able to answer the question on site type, to identify a portion of the physical landscape that they manage. This is not to say, however, that the forprofit sector is not involved in the local environment; it is simply not involved in the stewardship function of public lands in the same way as non-profit groups. Rondinelli and London (2003) describe firm-NGO relationships of differing intensities, with the most common being the "arm's length" relationship, which includes corporate donations and employee volunteerism. The survey shows that 18.5% of respondents listed corporate donations as one of their top three sources of funding, the third highest ranked funding source overall. Also, the involvement of corporate volunteers in large-scale one time park clean-up days and other events is quite common. Sustained environmental stewardship, however, is not generally a long-term function filled by these firms unless representatives function in a dual capacity of citizen and business leaders.

Groups are defined by more than whether they are public or private entities. Organizational culture, which can be understood in a limited way by analyzing missions and major programs, fundamentally contributes to the way a group "does business." Wilson notes (1989),

> "Every organization has a culture, that is, a persistent patterned way of thinking about the central tasks of and human relationships within an organization. Culture is to an organization what personality is to an individual. Like human culture generally, it is passed on from one generation to the next. It changes slowly, if at all" (91).

Based on the coding of open-ended reporting of missions and major programs, stewardship groups focus on improvement of environmental quality (22.5%), community development (39.2%), and environmental education (38.3%), showing that the groups have environmental and community values. Situating urban ecological stewardship within the chronology of the environmental movement provides an understanding of how these groups map onto waves of protectionism, conservationism, populist environmental advocacy, and environmental justice (see Figure 4). Generally, urban stewardship organizations are young, with over 90% founded since 1970. This is not surprising, given the rise in urban 'self-help' social movements during the 1960s and 1970s.

Reviewing the data respondent-by-respondent, organizations founded prior to 1960 included government entities like the National Parks Service and the Metropolitan Council of Governments. Comparing civil society and stewardship organizations overall shows that patterns are similar, reflecting the increase across all sectors in environmentalism. The mean founding date of all stewardship groups is late 1981. There is a marked rise in stewardship groups founded since 2000, which may continue to rise given that newer organizations might have been systematically under sampled from a parent population based on databases that are in some cases up to three years old. Further research on these newer organizations is needed.

5.4.2 Organizational Resources: Staff, Budget, Funding Source, and Information

An examination of organizational resources is useful for two reasons: 1) it helps to evaluate one dimension of the capacity of these stewardship organizations to pursue their missions, again framed by the HEF concept of critical resources, and 2) it reveals one layer in the stewardship network, the relationship between funders and recipients, and a capital flow in the HEF model. These resources are examined through questions on staff, budget, funding sources, and information.

Staff size is an important measure of the level of development and formality of an organization, and looking at staff size and community volunteer base together can give a sense of how an organization accomplishes its work and at what scale (see Figure 5). The stewardship groups are generally small in size, with 63.8% of all organizations and 80.7% of civil society organizations having fewer than ten full time staff. The number of organizations with zero full time staff is also notable, with many of the groups operating entirely on a volunteer basis. Groups with zero full time staff were not just the volunteer community groups as one might expect, but were evenly divided between formal nonprofits and informal groups.

Another surprising finding was the large number of groups with zero or less than ten community volunteers, as stewardship is popularly associated with high levels of volunteerism. There were seven civil society groups that reported having both zero full time staff and zero community volunteers, relying upon part time staff, part time volunteer staff, consultants, and contractors. These all-volunteer groups serve the community informally by creating public green space and beautifying neighborhoods, but they count members as the only participants in their programs rather than users of the site. A count of the latter would reveal broader impact more clearly.

Fisher and Green (2004) argue that staff capacity (among other endogenous resources) can be a barrier leading to disenfranchisement of civil society organizations and developing countries from international sustainability negotiations and politics. Particularly in large metropolitan areas, local political decisions also require time, resources, savvy, and lobbying, which should limit the ability of stewardship groups to participate. While some stewardship-only groups may not be interested in local politics *a priori*, they can become engaged when the sites that they manage are threatened, as was true in the 1990s during the closing, auction, bulldozing, and development of a number of community gardens in New York City (von Hassell 2002). In that case, full time staff was not the limiting factor, as these groups tended to rely upon volunteers working through a community organizing process and building coalitions with likeminded garden groups, using outsider tactics like protest and street theatre. In parallel, larger nonprofits like Trust for Public Land used insider tactics, including discussions with the city and the Attorney General and the buying up of auctioned garden sites. Community organizing around threatened gardens is beyond the scope of this paper, it is raised as one example of the way in which crises can politicize even previously non-political stewardship groups (a 'triggering event,' described in (Carmin and Hicks 2002)), at which point the interaction between resources and political participation becomes even more salient.

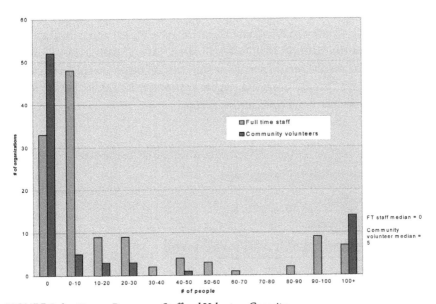

FIGURE 5.4   Human Resources: Staff and Volunteer Capacity

Budget can be considered one of a group's most fundamental resources (see Figure 6). Budget—along with volunteer staff and in kind donations—entirely determines the level of possible staffing and on the ground programs. Over 16% of the civil society organizations function with a budget of under $1,000/year, indicating a large, grassroots, under-resourced portion of the network.

In contrast, just one organization categorized as a local public agency (a public school environmental group), had a budget of under $1,000/year. These small budget groups include the site-specific stewardship groups, such as community garden groups, school garden groups, neighborhood park "friends of" groups, and environmental "clubs". The network is not entirely without financial resources, however, as over 64% of these organizations have budgets of larger than $100,000/year. The intermediate-sized nonprofit organizations with budgets of $100,000-$500,000 include citywide groups like the New Haven Land Trust and the Boston Toxics Action Center, as well as larger environmental education groups. Those with resources over $1 million include high profile citywide friends-of parks groups like the Pittsburgh Parks Conservancy, as well as nationally significant nonprofits (many of which were located in Washington, D.C.) like American Forests and the America the Beautiful Fund. Seventy-six percent of public agencies have budgets of over $100,000. The ten organizations with budgets over $5 million include the Parks and Recreation departments of these major cities, as well as some county agencies with responsibility for the metro area (e.g. County of Allegheny Department of Parks) and federal groups responsible for the National Mall in Washington, D.C. The diversity of groups even within the mantle of urban ecology stewardship helps to explain the wide range of budgets that are observed. Figure 6 shows the contrast between the budgets of civil society and government groups.

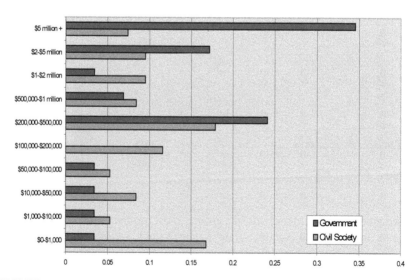

FIGURE 5.5   Percentage of Groups by Budget Category

Despite the available resources, 49% of groups in the survey identified "lack of funds" as the top barrier to the successful pursuit of their organizational missions (see Figure 7). The second highest barrier was "lack of staff" at 23%, which is at least partially a function of lack of funds. These responses were generated in response to an open question rather than picking a response from a list. Additional barriers include (in rank order): lack of time, bureaucratic barriers, lack of cooperation, and lack of political power. Moreover, respondents were asked if they agreed with the statement "this budget adequately serves my group's needs." Fifty-three percent of respondents disagreed (and 27% were neutral). Therefore, we can conclude that the current allocation of resources is not meeting the needs of the majority of urban ecology organizations. Whether it is an issue of absolute resources or allocation is not known, but it makes the need for leveraging resources all the more important. Indeed, the potential to leverage resource and pursue joint fundraising was one of the motivators behind the formation of the multi-city collaborative (the UEC) that supported the assessment discussed here.

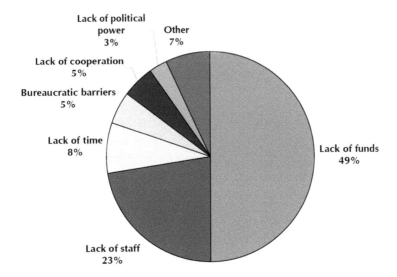

FIGURE 5.6   Top Identified Barriers to Achieving Mission

The question on funding sources asked respondents to select their top three funding sources (unranked); figure 8 shows the percent of all respondents that included each funding source in their top three. Unsurprisingly, municipal government (32.1%), state government (22.6%), and federal government (20.8%) were the top three sources of funding for public agencies. All other sources were ranked highly by no more than 11% of public agencies. Local foundations (42.7%) and private giving/membership (32.9%) are the top two sources for civil society organizations. It would have been useful to separate membership fees from private giving. Further confounding these responses was the separation of fees/program income from giving/membership. Despite these potential wording issues in the assessment tool, it is evident that more than 50% of stewardship groups rely on the financial support of individuals (through fees and donations) as one of their primary funders. All government funding sources combined were selected by 41.6% of respondents as being primary funders. The insufficient budgets and small staff sizes combined with a heavy reliance upon local foundations corroborate assessment research that small stewardship non-profits lack much-needed support for general operating expenses (Svendsen and Campbell 2005). While there is private foundation funding available to support program expenses, general operating resources are scarce, making organizational growth and sustainability a real challenge. Environmental stewardship organizations are also supported by the private choice of individuals through in-kind and volunteer support. Since they are less reliant on public funding, this contribution should be considered a "source" rather than a "sink" of human and social capital. They should be supported and used as conduits to affect environmental

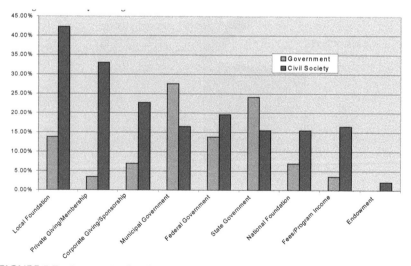

FIGURE 5.7   Primary Funding Sources

change, rather than ignored or reinvented, as some government-led programs tend to do (Burch and Grove 1993).

The HEF model categorizes information as a critical socioeconomic resource. Since the UEC was formed in part to support better information exchange amongst stewardship groups, the survey wanted to determine how easily stewardship groups can access information and "successful models" in their field.[4] Over 72% of all organizations and all civil society organizations agreed that they could access these models. This finding was surprising given the perceived programmatic redundancies and inefficiencies that can be observed amongst small, developing nonprofits. What, then, is the role for government and private foundations interested in supporting research, networking, and information clearinghouses? It seems to suggest that these agencies and funders could be encouraged to move away from the current model of 'technology transfer' and more towards one of capacity building through 'technology exchange.' The issue is less one of availability of technical information and more one of coproduction of knowledge (Fischer 2000). In this case, stewardship organizations reported that the primary resources they provided to community were: information (54%), hands-on training (41.5%) and volunteers (37.8%); see Figure 9. These data are used by groups internally to improve programs and services (58.5%), to satisfy funders' requests (54%) and to create legitimacy and a constituency.

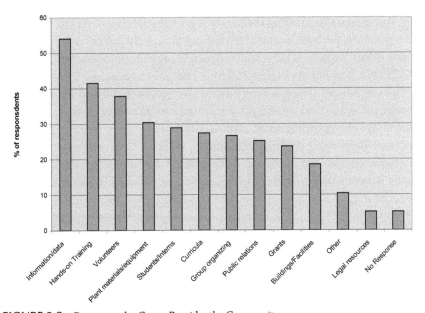

FIGURE 5.8   Resources the Group Provides the Community

5.4.3 Organizational Networks: Audience, Partnerships, Networking Strategies

Stewardship groups, like all organizations, have networks that connect them to other organizations and actors both vertically and horizontally. For instance, government agencies, funders, and intermediaries interact with stewardship groups by providing funding, technical assistance, information, as well as material resources (such as soil, tools, landscaping equipment, etc). The stewardship groups themselves interact horizontally with other stewards, coalitions, and advocacy nonprofits that share a common interest in urban ecology. Finally, stewardship groups interact directly with individual members, neighborhood residents, schoolchildren, and one-time and sustained volunteers. Groups were asked to describe their existing networks in both directions, in terms of audience and fellow stewardship groups. Determining which partners are considered critical to the functioning of these groups and what groups they would like to work with in the future was considered critical for network analysis.

Since the assessment was implemented in two rounds, with Boston and New Haven conducting outreach in late winter/early spring 2004 and the remaining cities conducting outreach in summer 2004, two different versions of one question were asked. For the first set, the question asked "what is the target audience of your programming?" and respondents were asked to choose all groups that apply. Participants conducting the survey reported confusion over the wording in this question, perhaps because stewardship groups do not consider partners or participants "audiences". Overall, civil society organizations selected: individuals (72.7%), community groups (63.6%), and public agencies (59%) as their top three audiences. The question's intent was reconsidered and its' phrasing reconfigured to ask "with what type of organizations does your group most often work?" Here the distribution of civil society organizations responses shifted away from individuals to other community groups (72%), schools (62.3%), and nonprofits (58.7%) as the top three selected. For public agencies, the top selected partners were schools (61.8%), community groups (61.8%), and nonprofits (61.8%)

By operationalizing the question of partnership in multiple ways, the assessment sought to get a better understanding of relatively who works with whom. Respondents were asked to rank other stewardship groups by the frequency with which they partner. The distribution of partners looked very similar between government respondents and civil society respondents. Both sets of groups ranked government groups as the stewardship group with which they most frequently partnered, (consistently/year round for 54% of civil society and 86% of public entities). Both groups tended to work a great deal with nonprofits, though civil society organizations had more interaction with individuals, and both worked infrequently with business groups. The distribution for just the civil society organizations is shown in figure 10. With the exception of the business sector, the majority of respondents reported partnering with all other stewardship groups frequently or consistently. This result could potentially be a function of the survey design and implementation. If anything, though, this question simply reinforces the lack of involvement on the for-profit sector in this capacity. It also reiterates the fact that government agencies (including municipal, state, and federal parks department as well as less obvious groups like water-based or agricultural agencies) are important stewards.

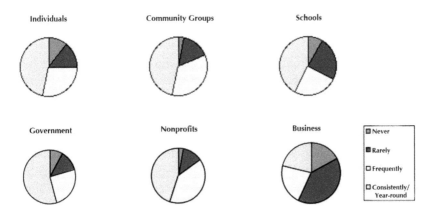

FIGURE 5.9   Frequency with which Civil Society Organizations Partner

The assessment asked respondents to identify and rank up to six organizations or individuals that were "critical to their work" currently. They were also asked to rank the top six individuals or groups with whom they would like to work with, in the future but are not currently. These two questions, taken together, move towards an understanding of the beginnings of a network—though not as loosely defined as the community of common values that Batterbury (2003) describes.

Comparing these responses side-by-side allows us to understand where this network currently stands and the direction in which it may evolve. Current organizations mirrored the responses to the stewardship partner questions, with city agencies and non-profits being the highest ranked responses. Of the non-profits listed, 19 were specifically environmental nonprofits, three were "cultural" non-profits, and one was a healthcare nonprofit. Of the city agencies, 15 were specifically referring to parks departments of the various cities, which continue to play critical roles in urban environmental stewardship. Other named agencies include health, environmental services, planning, and urban forestry departments. Finally, of the 12 organizations listing state agencies as key partners, 10 of these were state natural resource departments.

For the future, respondents ranked highest a variety of environmental groups, government agencies, and research groups. The high ranking of research as a priority area is surprising, and perhaps suggests the potential for community based or participatory research that takes advantage of the existing close relationship between government agencies and local stewards. Also notable is the rather high rank of business groups; it seems that the stewardship groups are aware of this gap in their network. Both grouped lists are shown in Table 2.

Beyond knowing who is in the network or who groups would like to have in the network, the assessment sought to find out what particular networking *strategies* organizations used to connect with other groups. Here, there was little variation between civil society and government actors. The most commonly used strategies by civil society organizations were: attending local community meetings (76.9%), generating press (71.4%), and participating in regional coalition group (67%). The high response for regional coalition was surprising, given a common perception of a lack of regional information-sharing and formal collaborative entities. Perhaps this reflects some ambiguity of the meaning of the word regional. The partners of the UEC and others are interested in using intermetropolitan coalition in order to affect change in individual cities. Other common strategies listed were attending national conferences (61.6%) and participating in citywide coalitions (57.1%). For government groups, the top three strategies were public-private partnerships (83.3%), participating in regional coalition groups (76.7%) and 73.3% said they attend local community meetings and generate press. Since public-private partnerships did not rank highly on the strategies of civil society organizations, it remains a question as to what are the groups with whom these government actors are partnering. The lowest ranked strategy in both cases was "participate in list servs", reflecting the reliance on face-to-face rather than virtual collaboration. When urban groups can physically meet, they seem to prefer that to virtual communication.

TABLE 5.2　Top Ranked Current and Future Partners

Top Ranked Current Organization	Count	%
City Agencies	34	30.63%
Non-profits	23	20.72%
State Agencies	12	10.81%
Community Groups	9	8.11%
School Groups	8	7.21%
Federal Agencies	7	6.31%
Business/Industry Groups	5	4.50%
Grantmakers (local)	5	4.50%
Research Groups	3	2.70%
Regional Agencies	2	1.80%
City Policymarkers	1	0.90%
State Policymarkers	1	0.90%
Legal Groups	1	0.90%
TOTAL	**111**	**100%**
no response	24	

Top Ranked Future Partners	Count	%
Environmental Groups	22	26.19%
Government Agencies	21	25%
City	*12*	
State	*1*	
Federal	*5*	
None Specified	*3*	
Research Group	12	14.29%
Business/Industry Groups	10	11.90%
Neighborhood Groups	6	7.14%
City-Neighborhood Planning Groups	2	2.38%
Religious Groups	2	2.38%
School Groups	2	2.38%
Sports Groups	1	1.19%
Funding Groups	1	1.19%
Celebrity Groups	1	1.19%
Preservation Groups	1	1.19%
African American Groups	1	1.19%
Volunteer Groups	1	1.19%
Youth Groups	1	1.19%
TOTAL	**84**	**100%**
No Response	51	

5.4.4 Biophysical & Social Impacts: Scale of Service, Neighborhood, Site Type, Land Jurisdiction

The final aspect from the UEC assessment that is considered here is how these groups' activities play out across the space of the urban landscape in terms of scale of service delivery and areas of stewardship work by neighborhood and site type. The HEF model includes biophysical resources as a major component of the human ecosystem. While this survey did not involve any physical land assessment or inventory of sites, it does capture where and how these groups organize on the landscape to demonstrate where the overlaps and gaps between groups are, which is a first step to establishing the link between organizations and physical resources.

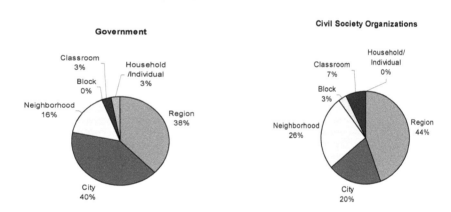

FIGURE 5.10   Scale of Service Delivery

A high number of groups indicated that they work across regions. While this was intended to mean metropolitan areas, upon reviewing the group's missions and self-descriptions, it may have been selected for different reasons. Many of the Washington, D.C. based groups selected "region", perhaps because they thought it better defined the District than did the term "city." Second, a number of watershed, stream, or other groups that were operating on an ecological rather than a political scale, were a selected region because of its more flexible usage. Civil society organizations comprise the strong majority of groups working at the neighborhood, block, and classroom scales, with most government agencies working city and region-wide. This pattern fits with our intuition about the civil society groups, given that most of them are small in terms of staff and resources; many groups have an intensely local focus. Why is it, then, that there is a perceived chasm between environmental interests and community development interests both on the ground in urban neighborhoods and in the academic literature? (Campbell 1996; Evans 2002). Is there a greater role for stewardship of the environment in the stabilization and development of neighborhoods? Research has documented aspects of this function, particularly in terms of open space's impact on property values and the importance of planning for active living to promote healthy communities (Harnik 2000; Frumkin 2003). But there is a need for further exploration of the links between the social act of stewardship/caring for the environment and public health, crime, and social cohesion. Findings from this assessment suggest that the stewardship motivations conflate improving the physical site, inspiring people to positive action and impacting the overall neighborhood. Figure 12 shows the social and environmental impacts that groups reported achieving.

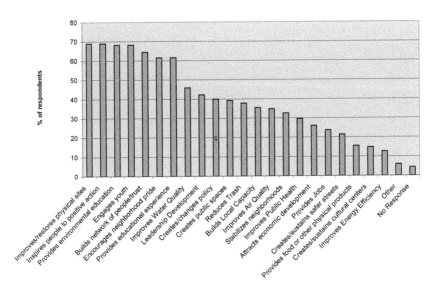

FIGURE 5.11   Social and Environment Impacts

While scale explains one dimension of group influence and describes one dimension of group capacity, geographically locating stewardship "spheres of influence" is suggested as a useful tool for the ecological planner, manager, designer and community organizer. For example, as the ecological planner tries to create recreation and nature corridors such as greenways, it is necessary to know both where the potential users and maintainers of these sites are. The community organizer needs to know where the clusters of high and low stewardship activity are for the purposes of focusing her outreach efforts or coalition building, for example. Groups were asked to identify both the neighborhood in which they work as well as the physical boundaries of where the group works (down to the block and street level). Neighborhood information for the New York City groups was geocoded and made into a sample map shown in figure 13. With further refinement at the neighborhood scale, this map could be developed for long-term use by urban environmental managers.

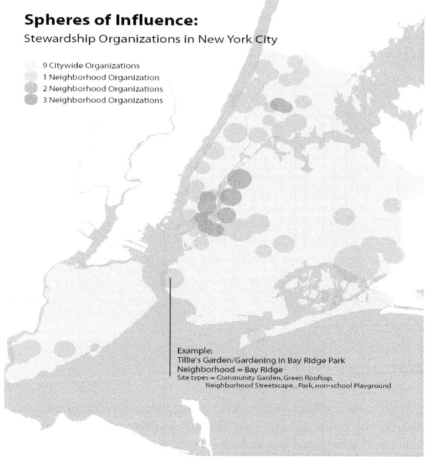

FIGURE 5.12   Spheres of Influence Map. Source: E. Svendsen & L. Campbell, Urban Ecology Collaborative and C. Spielman, Community Mapping Assistance Project, 2004.

Within each city and neighborhood, there exists a diversity of site types. Respondents were asked to select from a list of 36 site types that were developed jointly by the UEC Research Sub-Committee to represent the range of sub-neighborhood site types within the Forest Opportunity Spectrum (Raciti et al. 2006). Overall, the top ranked sites were park, watershed, protected/natural site, stream/river/canal, and waterfront. Every site type was selected by no fewer than nine respondents. The thirty-six site types can be categorized into four general categories. Designated open space, including both recreational space like playgrounds and recreation parks as well as ecological space like natural protected areas, is the most frequently stewarded site type (34.1%). Water related sites (26.8%) include the expected: streams, waterfronts, estuaries, as well as the less conventional: underground streams and sewersheds. Built environment (20.5%) includes any green space on buildings or building sites, including green rooftops and courtyards, but also vacant lots and brownfields. Neighborhood streetscape (18.6%) includes all of the sites that are not on dedicated open space or building parcels, so this includes street trees and planters, but also highway medians, public right of ways, street ends, and traffic islands. Figure 14 shows the ranking of all the site types that were selected.

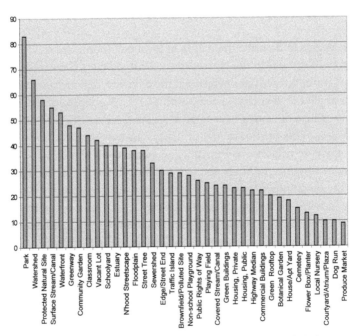

FIGURE 5.13   Number of Organizations Working on Site Types in Rank Order.

The final aspect to consider related to perceived impact is the jurisdiction of the various site types. Given the distribution of site types that includes the built environment and streetscape in substantial numbers, it is clear that stewardship is not just occurring on officially designated and publicly managed open space. In total, publicly held property does comprise the majority of sites on which all stewardship groups work at 57.5% (56.6% for just civil society orgs). Municipal government is the most common landowner of these sites, followed by state, and then federal government. Public managers must heed this presence of independent stewardship groups acting on public lands; for, as many have observed the design, use and meaning of public space is constantly challenged in the modern city (Jacobs 1961; Cranz 1982; Jackson 1984; Rosenzweig and Blackmar 1992).

The remainder of sites is divided almost evenly between individually owned land (15%), nonprofit owned land (15%), and business owned land (12%). Managing the city as an ecosystem would require coordinated action across parcels with different management objectives and stewardship groups. Inventorying and making publicly available information on site jurisdiction is one critical first step, even independent of further research on organizations.

5.5   CONCLUSION

This paper begins to describe the nature of local environmental stewardship in large metropolitan areas in the Northeastern United States. Stewards are a mix of a few, larger public agencies operating at the citywide, regional and state scales and many smaller civil society actors, both 501(c) 3 nonprofits and informal community groups operating in ecological regions, across cities, and in specific neighborhoods. This organizational diversity can be viewed as both a source of social capital, in response to Putnam, and evidence of vibrant local environmentalism, in response to Shellenberger and Nordhaus. Public interest in the quality of the environment may in fact be on the rise, in response to Greenberg, but is nested within a larger context of quality of life issues. Finally, Harvey's notion that urban environmental groups are fragmented and inefficient is unresolved. The extent to which these groups will become further fragmented within specific spheres of influence or begin to develop organizational mechanisms in which to partner is unknown at this time. There is a strong underlying assumption made by this paper that without the introduction of a perceived crisis or risk, the only way to harness the capacity of stewardship groups is through deliberate multi-scaled, capacity building networks.

The assessment discovered a dynamic social network of organizations within cities with a reserve of social capital and expertise that could be better utilized. Although not the primary land owner of the sites on which they work, stewardship groups take responsibility for a wide variety of land use types. Outputs include the delivery of public programs as well as site maintenance. Most of the groups work in collaboration with government managers but operate on staffs of zero or fewer than ten, with small cohorts of community volunteers (and potentially large numbers of 'site users'). Resources are scarce and inconsistent, making it a challenge for groups to grow beyond their current capacity to develop long-term programs critical to education and management. This creates an impression of fragmentation which may not be legitimate given that certain events have the potential to unite groups across place and between scales. Stewardship networks are rather self-contained and while the business sector and legal groups are present, they are not sufficient given the critical resources that these groups can provide. This presents a challenge both to stewardship groups themselves (in terms of their own sustainability) and to planners and land managers that attempt to work with these groups. Research partnerships and shared governance structures are two potential means by which this network could be expanded (Durant 2004).

More comprehensive research of these groups is needed to be able to ask second order-questions, like the relationship between ideologies, management type, resources, strategies, and outcomes. Further research is also needed to explore the full breadth and complexity of the stewardship network. This study is a first attempt to understand groups with some affiliation to environmental umbrella organizations, but we recognize that there is a much larger universe of civil society groups for which environmental concerns are nested within other priorities (e.g. green career groups, faith based groups, youth oriented groups). An understanding of the full stewardship network will need to be cultivated in order to support stewards' work in restoring and revitalizing urban ecosystems and human communities.

FOOTNOTES

1. The New York City assessment was conducted in partnership with New York University's Wallerstein Collaborative. A special thanks to Dr. Mary Leou and her graduate assistant, Lisa Babcock.
2. Partnerships for Parks and NYC Dept. of Parks and Recreation GreenThumb Program.

3. The Business Improvement Districts (BIDs) are an important and engaged stewardship group in the City of New York. However they were not included in this limited sample but are strongly suggested for inclusion in future research.
4. The survey also asked a question on access to scientific information, but response rate was extremely low and respondents had difficulty ranking the various choices, so that question is not considered here.

REFERENCES

1. Agyeman and Evans, J. and. Evans, T.. 2003. "Toward Just Sustainability in Urban Communities: Building Equity Rights with Sustainability Solutions." The Annals of the American Academy 590: 35-53.
2. Alberti, M. 2005. "The Effects of Urban Patterns on Ecosystem Function." International Regional Science Review 28(2): 168-192.
3. Batterbury, S. 2003. "Environmental Activism and Social Networks: Campaigning for Bicycles and Alternative Transport in West London." The Annals of the American Academy, 590: 150-169.
4. Andrews and Edwards, K. and Edwarads, B.. 2005. "The Structure of Local Environmentalism." Mobilization 10 (2).
5. Brachman, L. 2003. Three Case Studies on The Roles of Community-Based Organizations in Brownfields and Other Vacant Property Redevelopment: Barriers, Strategies and Key Success Factors. Working Paper, October. L. I. o. L. Policy, Lincoln Institute of Land Policy.
6. Burch, W. R., Jr. and Grove, J. M.. 1993. " People, Trees and Participation on the Urban Frontier." Unasylva 44:: 19-27.
7. Campbell, S. 1996. Green cities, growing cities, just cities? Readings in Planning Theory, 2nd Edition, 2003. S. Campbell and S. S. Fainstein. Oxford, London.
8. Carmin, J., Hicks, B. and Beckmann, A. 2003. "Leveraging Local Action: Grassroots Initiatives and Transnational Collaboration in the Formation of the White Carpathian Euroregion." International Sociology 18(4): 703-705.
9. Carmin, J. and Hicks, B. 2002. "International Triggering Events, Transnational Networks and the Development of the Czech and Polish Environmental Movements." Mobilization 7(3): 305-324.
10. Coban, A. 2003. "Community-based Ecological Resistance: The Bergama Movement in Turkey." Environmental Politics 13(2): 438-460.
11. Cranz, G. 1982. The Politics of Park Design: A History of Urban Parks in America. Cambridge, MIT Press.
12. Dalton, R. J., Recchia, S., et al. 2003. "The Environmental Movement and the Modes of Political Action." Comparative Political Studies 36(7): 743-711.
13. Dalton, S. E. 2001. The Gwynns Falls Watershed: A Case Study of Public and Non-Profit Sector Behavior in Natural Resource Management. Baltimore, MD, Johns Hopkins University Press: 152 pgs.
14. Durant, R., Daniel, F., Fiorino, J., and O'Leary, R.. 2004. Environmental governance reconsidered: challenges, choices, and opportunities. Cambridge, MIT Press.
15. Edwards, B. and Foley, M. W.. 1998. "Civil Society and Social Capital Beyond Putnam." American Behavioral Scientist 42: 124-139.

16. Evans, P. 2002. Political Strategies for More Livable Cities. Livable Cities. Berkeley, University of California: 222-246.
17. Fischer, F. 2000. The Return of the Particular - Scientific Inquiry and Local Knowledge in Postpositivist Perspective. Citizens, Experts and the Environment: The Politics of Local Knowledge. Durham, Duke University Press: 68-85.
18. Fisher, D. and. Green, J. F. 2004. "Understanding Disenfranchisement: Civil Society and Developing Countries' Influence and Participation in Global Governance for Sustainable Development." Global Environmental Politics 4(3): 65-85.
19. Fisher, D. R. 2004. National Governance and the Global Climate Change Regime. New York, Rowan & Littlefield Publishers, Inc.
20. Frumkin, H. 2003. "Healthy Places: Exploring the Evidence." American Journal of Public Health 93(9): 1451-1454.
21. Greenberg, M. 2005. "Environmental Protection as a US National Government Priority: Analysis of Six Annual Public Opinion Surveys, 1999–2004." Journal of Environmental Planning and Management 48(5).
22. Grove, J. M. and Burch, W. R. J. 1997. "A Social Ecology Approach and Application of Urban Ecosystem and Landscape Analyses: A Case Study of Baltimore, Maryland." Urban Ecosystems 1: 259-75.
23. Grove, J. M., Hinson, K., and Northrop, R.. 2002. Education, Social Ecology, and Urban Ecosystems, with examples from Baltimore, Maryland. Understanding Urban Ecosystems: a new frontier for science and education. A. R. Berkowitz, C. H. Nilon and K. S. Hollweg. New York, SpringerVerlag: 167-186.
24. Harnik, P. 2000. Inside City Parks. Washington DC, Urban Land Institute.
25. Harvey, D. 1999. The Environment of Justice. Living with Nature: Environmental Politics as Cultural Discourse. F. a. M. H. Fischer. Oxford, Oxford University Press: 153-185 pgs.
26. Jackson, J. B. 1984. Discovering the Vernacular Landscape. New Haven, Yale University.
27. Jacobs, J. 1961. The Death and Life of Great American Cities. New York, Random House.
28. Keck, M. E. and. Sikkink, K.. 1998. Activists Beyond Borders: Advocacy Networks in International Politics. Ithaca, Cornell University Press.
29. Machlis , G. E., Force, J. E., et al. 1997. "The Human Ecosystem Part I: The Human Ecosystem as an Organizing Concept in Ecosystem Management,." Society and Natural Resources. 10: 347-367.
30. McCreary, S., Gamman, J., Brooks, B., Whitman, L., Bryson, R., Fuller, B., McInerny, A., and Glazer, R. 2001. "Applying a mediated negotiation framework to integrated coastal zone management." Coastal Management 29: 183-216.
31. Millennium Assessment, U. N. 2005. Investing in Development: A Practical Plan to Achieve the Millennium Development Goals. Report to UN Secretary General. J. Sachs, United Nations: 94 pgs.
32. Niemela, J. 1999. "Ecology and Urban Planning." Biodiversity and Conservation 8: 119-131.
33. O'Rourke, D. and. Gregg, M.. 2003. "Community Environmental Policing: Assessing New Strategies of Public Participation in Environmental Regulation." Journal of Policy Analysis and Management 22(3): 383-414.
34. Pickett, S. T. A., Burch, W. R. J., et al. 1997. "A Conceptual Framework for the Human Ecosystems in Urban Areas." Urban Ecosystems 1: 185-199.
35. Program, H. E. 2002. About HEP. Accessed through <http://www.harborestuary.org/> (15 April 2005).

36. Putnam, R. 2000. Bowling Alone. NY, Simon and Schuster.
37. Raciti, S., Galvin, M. F., Grove, J. M, O'Neil-Dunne, J. P. M., Todd, A., and Clagett, S. 2006. Urban Tree Canopy Goal Setting: A Guide for Chesapeake Bay Communities, United States Department of Agriculture, Forest Service, Northeastern State & Private Forestry, Chesapeake Bay Program Office, Annapolis, Md.
38. Redman, C. L. 1999. "Human Dimensions of Ecosystem Studies." Ecosystems 2: 296-298.
39. Rosenzweig, R. and Blackmar, E.. 1992. The Park and the People: A Social History of Central Park. New York, Cornell University Press.
40. Sampson, R. J. and. Raudenbush, S. W.. 1999. "Systematic Social Observation of Public Spaces: A New Look at Disorder in Urban Neighborhoods." American Journal of Sociology 105(3).
41. Shellenberger, M. and Nordhaus, T.. 2004. The Death of Environmentalism: Global Warming Politics in a Post Environmental World, <http://www.thebreakthrough.org/images/Death_of_Environmentalism.pdf> (1 May 2005).
42. Sirianni, C. and Friedland, L. 2001. Civic innovation in America : community empowerment, public policy, and the movement for civic renewal. Berkeley, University of California Press.
43. Skocpol, T. and. Fiorina, M. P.. 1999. Civic Engagement in American Democracy. Washington DC, Brookings Institute.
44. Svendsen, E. and Campbell, L.. 2005. The Living Memorials Project: Year 1 Social and Site Assessment. GTR-NE-3333. Washington, DC, USDA Forest Service.
45. UEC. 2004. "Urban Ecology Collaborative." accessed through <http://www.urbanecology-collaborative.org> (5 May 2005).
46. von Hassell, M. 2002. The Struggle for Eden : Community Gardens in New York City, Bergin & Garvey.
47. Wapner, P. 1995. "Politics Beyond the State: Environmental Activism and World Civic Politics." World Politics 47(3): 311-340.
48. Weber, E. P. 2000. "A New Vanguard for the Environment: Grass-Roots Ecosystem Management as a New Environmental Movement." Society and Natural Resources 13(3): 237-259.
49. Westphal, L. M. 1993. Why Trees? Urban forestry volunteers values and motivations. General Technical Report NC-1633. P. H. Gobster. St. Paul MN, USDA Forest Service North Central Research Station: 19-23.
50. Wilson, J. Q. 1989. Bureaucracy: What Government Agencies Do and Why They Do It. New York, Basic Books.
51. Wondolleck and Yaffee, J. M., and Yaffee, S.L.. 2000. Making Collaboration Work-Lessons from Innovation in Natural Resources Management. Washington DC, Island Press.

CHAPTER 6

Planning Office and Community Influence on Land-Use Decisions Intended to Benefit the Low-Income: Welcome to Chicago

Yan Dominic Searcy

6.1   PURPOSE OF THE STUDY

Much of the literature on poverty focuses on federal level initiatives as a response to urban poverty, seeming to ignore local level political processes. Researchers have tended to accept Peterson's [1] contention that cities will not redistribute their own resources and that actual redistribution, utilizing monies from taxes from upper income groups to support initiatives that benefit lower income groups, can only occur with federal assistance [2, 3]. As a result, scarce attention has been given to policy efforts that utilize local resources that are intended to address urban poverty. A few researchers, mainly urban planners, suggested that it is possible to address poverty at the local level but offered little substantiation of what could be done or what was done [4–7].

Urban planning researchers Mayer [4], Pierre [5], and Wong [6] suggest that planning offices and community involvement can influence local policy decisions so that they benefit low-income urban residents. The researchers also suggest political conditions which may allow for policy innovations that can be couched in technical and bureaucratic processes that may yield potential benefits for low-income residents. However, missing from the literature are examinations of instances of planning office and community influence on the use of local resources to address urban poverty.

Broadly, this study explores a local policy effort that utilized land as a local resource with the stated intention to address urban and neighborhood-based poverty. Specifically, the study (1) examines to what degree and under what conditions the Chicago Department of Urban Planning and communities influenced decision making regarding land use for low-income residents through an analysis of two neighborhood areas in Chicago from 1990–1997. The hypothesis proffered is that community-based political conflict would be the factor that influenced the degree to which land-use decisions would benefit low-income residents. This research follows Clark's [7] suggestion not only to consider model cities but also to study how particular policies are shaped and made.

6.2 METHODOLOGY

Contact between planning offices and communities generally occurs during neighborhood revitalization or redevelopment efforts. Although urban poverty is neighborhood-based, revitalizing urban neighborhoods is not implicitly an exercise in addressing poverty for two major reasons. First, not all neighborhoods undergoing revitalization are low-income. Second and most importantly, revitalization aims to eliminate or prevent urban blight but not to ameliorate poverty. As the literature on gentrification indicates, those neighborhoods with low-income residents that are undergoing redevelopment often simultaneously experience displacing the poor [8, 9]. Responding to displacement of the poor, community organizations have attempted to influence policies to balance redevelopment by advocating for low-income housing for the poor in neighborhood areas undergoing redevelopment. The intersection where both planners and communities come together during revitalization efforts is the neighborhood.

To explore urban planning office and community influence on land-use decision making, two Chicago neighborhood areas undergoing revitalization, Woodlawn and Kenwood, were studied. The Chicago Department of Planning and Development was active in both neighborhoods and produced redevelopment plans and conservation plans for the respective neighborhoods. The study areas also provided instances where the Department approved land uses that intended to benefit the low-income. Both neighborhoods were also home to community organizations that attempted to influence redevelopment efforts and land-use decision making.

A qualitative study of Woodlawn and Kenwood in neighborhood areas in Chicago is employed to address the research questions. Primarily because of the exploratory nature of the research and the inability to manipulate variables, a qualitative analysis is employed as an attempt to examine the dynamics and the degree of Department of Planning and Development (referred to here as the "Department") and community influence on land use. Influence of the Department is operationalized as departmental initiated land utilization that directly benefits the low-income. Community influence as utilized in this study is limited to referring to neighborhood-based organizations that utilized land to produce units of low-income housing. Merely focusing on the ability to impact outcomes does not yield an authentic assessment of influence. Influence is most accurately measured through an assessment of outcomes. That is, influence is best discerned through assessing the degree to which low-income units were produced.

To avoid selection bias which may predetermine the outcome of the study, the neighborhood areas were selected because of the presence of the independent variables, a planning office (the Department) and community organizations, rather than incidences of the production of low-income units [10]. The two neighborhood areas Kenwood and Woodlawn manifest the presence of both an urban planning office and community organizations. They were selected for this reason, not because they manifest land-use decisions that have resulted in the development of units of low-income housing. Doing so would have led to selection bias.

Though the use of only two study areas may serve to limit the generalizability of the results, the limited focus was intended to lead to more detailed, robust research than research that involves multiple case studies. Chicago is studied because as the third largest city in the United States its dynamics are arguably similar to a number of major US cities and therefore may lead to generalizability to cities such as New York, Cleveland, and Milwaukee that have had to address urban decline and revitalization demands. Limiting the generalizability of the study, however, is the uniqueness of Chicago political life that reflects the persistence of machine governance combined with elements of regime politics. One recent study labeled Chicago as reflecting a "mayor-centered neoclientelism."

[11] At the outset of the study period from 1990 to 1997 the presiding mayor Richard M. Daley (a son of long-term machine mayor Richard J. Daley) was one year into his 22-year tenure as the city's chief.

The time period from 1990 to 1997 reflects a need to address urban political dynamics that have received little qualitative or quantitative study and include the emergence of what Clark [7] labelled as a new political culture. The researcher's proximity and access to pertinent research resources are also factors that influenced the selection of Chicago as the focus of the study.

In this study, community is operationalized as nonprofit neighborhood-based organizations that serve and advocate for residents of a neighborhood. A problematic in discussions of community influence is the concept of "community." A caution is issued, therefore, that community influence as utilized in this study is limited to referring to the neighborhood-based organizations that were able to produce units of low-income housing. It can be argued that community organizations that may not have near unanimous support from neighborhood residents do not represent community-wide influence. Community, therefore, may only represent those members who are the most organized, not the most numerous.

The study utilizes the federal definition of low-income family incomes that are 65 percent or less than the Primary Metropolitan Statistical Area (PMSA) median income. During the study period the Chicago PMSA median income for a family of four was $41,745. Low-income families, therefore, would have incomes at or below $27,134. Housing structures that are considered low-income must not exceed 33 percent of the family income. Based on federal standards, an affordable rent or mortgage for a family of four should cost no more than $747. In order to give context to understanding redevelopment initiatives during the study period from 1990 to 1997, a detailed history of Woodlawn and Kenwood is provided below.

6.3   THE NEIGHBORHOODS

Through a series of transitions and a process of divestment Woodlawn and Kenwood became like fossils, skeletal and lifeless, but left with the imprint of what were once vibrant and thriving neighborhoods. By the late 1980s, both neighborhoods began revitalization efforts.

6.3.1   Woodlawn

Woodlawn lies approximately eight miles south of downtown Chicago.

The area's population peaked in the 1950s with 80,699 residents. The 1950s also marked the beginning of the transition of the German and Irish neighborhood to one almost exclusively Black. Though the 1960s marked Woodlawn as a figuratively fiery period of community activism, the 1970s was literally a fiery period for Woodlawn. The neighborhood experienced an unusual amount of fires. In 1970, there were 1,600 reported fires within the eastern section of Woodlawn [12].

The 1980s delivered Woodlawn another blow with increasing numbers of vacant lots, abandoned buildings, and increasing violence as crack cocaine changed the dynamics of street and gang crime. The neighborhood became a lab for poverty researchers and journalists [13, 14]. Between 1970 and 1980 Woodlawn suffered a loss of 6,477 units of housing and 32 percent of its already diminished population (36,323 residents in 1980) (Community Development Report 1992).

By 1990, over 40 percent of the eastern section of Woodlawn was vacant while slightly over 25 percent (1,169 vacant lots) of the entire neighborhood area was vacant (Community Development Report 1992). Woodlawn businesses numbered about 100. Thirty years before, the number was nearly 800. In 1990, over 53 percent of Woodlawn households made less than $15,000.

6.3.2   Kenwood

Kenwood, as referred to in this study, is actually the agglomeration of two contiguous neighborhood areas (North Kenwood and Oakland) whose histories are closely related. Residents and the Department consider the area to be one neighborhood, North Kenwood-Oakland, despite being municipally identified as two distinct neighborhood areas. Below, a brief history of both Kenwood and Oakland is provided in order to provide understanding of the dynamics that led to the decline of the neighborhoods and also to identify how they came to be considered as one neighborhood area.

6.3.3   North Kenwood

Kenwood is located approximately four and one-half miles south of Downtown Chicago and borders Lake Michigan to the east. In 1950, Kenwood was 10 percent Black. By 1960 the area was 84 percent Black. In 1956 the southern, more affluent and white end of Kenwood was designated as a conservation area and annexed as part of the Hyde Park-Kenwood Conservation Area (HPKCA). Implementation of the HPKCA plan brought massive demolition and divestment in the northern section of the neighborhood (Community Fact Book 1995).

During the 1970s, the area continued to suffer from divestment. The population that remained, 26,908 (a 36 percent drop in ten years), experienced increasing numbers of arson and abandonment, dilapidated housing ignored by absentee landlords, and increasing crime and violence. The 1980s greeted the neighborhood with a population drop to 22,000 and a continuation of the trends of abandonment and dilapidation of the 1970s. By 1990 the overall population of Kenwood had decreased to 18,000. The racial composition became 97 percent Black. Approximately 37 percent of its residents made less than $15,000.

6.3.4 Oakland

By the 1950s, the population of the area swelled to more than 24,000 (up from 15,000 in 1930) as a result of the growth of the Black population in the area. To address housing shortages, new construction was of public housing. Several housing projects were built in the area, concentrating low-income residents [12]. By 1960, the area was 98 percent Black. Toward the end of the 1960s, an urban renewal project razed one-fourth of the housing in the area and left the land vacant for nearly two decades.

By 1970, the area manifested the qualities of a slum, dilapidated and vacant housing, high concentrations of public housing, and high crime rates. In 1980, 71 percent of the total housing units in Oakland were publicly assisted. The 1980s, however, marked the beginning of redevelopment in Oakland as a public housing project was converted to mixed-income housing. Though redevelopment began, two of the five census tracts in Oakland ranked among the lowest income areas in the United States in 1990 as family median income was recorded to be below $5,400 for both tracts. In 1990, the population decreased to 8,197. Mean household income was $10,849. Slightly over 72 percent of Oakland's population was in poverty.

6.4   HANDS TO THE CLAY: INFLUENCE

Land-use decision making involves more than planners and community organizations. The role of the mayor and alderman cannot be marginalized. Therefore,
a brief discussion of the influence of the mayor and aldermen is provided before
Departmental and community organization influence is addressed.

6.4.1   Mayoral Influence

Redevelopment must have the mayor's direct sanction. It is nearly impossible
that any redevelopment could occur in Chicago without mayoral approval.
Through appointment powers, the mayor has the ability to influence the land-
use decision making process. The mayor appoints

  (i)   the commissioner of the Department,
  (ii)   the commissioner of the Department of Housing (as well as other city
         posts),
  (iii) members to the Urban Renewal Board,
  (iv)   members to the Community Development Commission,
  (v)   members to the Chicago Plan Commission on which the mayor sits,
  (vi)   members to the Neighborhood Planning Council,
  (vii) members to the Conservation Community Council.

The mayor also can remove appointments at his discretion. By virtue of his
appointment powers, the mayor guarantees that his development interests will
be primary.

Although the mayor may not produce actual development plans, his desires
are manifested in the activities or lack of activities of the Department. Mayoral
influence is also displayed as a budgetary issue as the mayor may allocate or fail
to allocate funds for particular Departmental initiatives. Corporate funding to
the Department from 1990 to 1991 reflected an over 30 percent increase. The
amount remained relatively consistent through 1997.

6.4.2   Aldermanic Influence

An alderman in this study can best be understood as a mayor of a ward/community.
The decisions that are made within the boundaries of the ward must be approved

by the alderman. Related to land use, the alderman has the ability to "hold" city owned land. That is, Chicago Aldermen can restrict the sale of city-owned properties within their wards. According to a planning official, "if the alderman does not push (a development plan), it does not matter who screams and hollers."

6.4.3 Planning Office Influence: A developer Driven Planning Office

In neighborhood redevelopment efforts, the Department is dispatched by the mayor and by aldermen. Developer interests, however, drive the planning process. As one Department official related, the Department is not composed of planning renegades who have the autonomy to canvas the city looking for areas to redevelop and then dictate the plans.

Although acknowledging that the mayor has a significant degree of power over the urban planning process, the official underscored the mayor's limitations, "the mayor can do what he wants, but (he) cannot do it without depending on developers." The same official added, "you cannot force a developer to come; the planning office cannot work on long term (development) and do land assemblage." The Department's residential planning goal is to assist the housing market by creating incentives for investment and then allowing market forces to direct the process. As the market sustains itself, the Department suspends its activities in a particular area.

The Department attempts to attract development with a number of joint tools: (1) studies to determine whether land use is to be commercial or residential or to change land use; (2) acquisition of property for tax reactivation by the Department Of Housing (DOH); (3) suggestions that land parcels are appropriate for housing based on the findings of preliminary studies; (4) composing redevelopment plans; and (5) composing conservation plans. The Department provided land surveys and zoning data, coordinated meetings, and carried out the overall administration of development. These activities represent the technical, professional process identified in the planning literature (Rabinovitz 1969; [5, 15]). Though the literature suggests that through the planning process, a planning office can influence land-use decisions intended to benefit the low-income, Chicago's Department focused on physical planning issues, not on issues of social equity.

In Kenwood, the Department saw the opportunity to "create something worthwhile" because vacant land was plentiful. City planners saw the redevelopment of Kenwood as potential to make a significant contribution to urban redevelopment. A planning official related,

> redevelopment was something that could be done and (land) was there and it was available to be done. The mayor did not hold it up… at the time there was no real politics in the Department. Some people think that underhanded stuff was going on but this is not always the case.

6.5   THE DECISION MAKING PROCESS

6.5.1   In Woodlawn

Woodlawn community redevelopment was initiated largely by a request from the Woodlawn Preservation and Investment Corporation (WPIC) which had been working on a number of development projects. Rev. Arthur Brazier, the chairman of the board of WPIC, attended to garnering city approval of the WPIC produced redevelopment plans. Brazier was a long-time community activist in Woodlawn whose history was tied to pastoring a Woodlawn storefront church and cofounding the seminal community organization The Woodlawn Organization (TWO) in the 1960s. By 1990, Brazier's storefront church had swelled to a membership of over 10,000 and was soon to begin building a multimillion dollar church on one full block in Woodlawn (completed in 1992). Also by 1990, TWO had grown into a multiservice organization that managed a multimillion dollar budget. Brazier became the central decision maker for three community organizations in the neighborhood: WPIC as chairman of the board, TWO as the organization's cofounder and former chairman of the board, and the Fund as chairman of the board.

According to a Departmental official who is also a Woodlawn resident, The Fund was created as a quasipublic development organization that was given oversight over development in the area (by the Department). The Fund's support from City Hall and its resulting influence is manifested in the organization's de facto control of Woodlawn redevelopment plans. That control manifested itself in the Fund securing the ability to be both a developer and to award development grants. The Fund and WPIC essentially became the "community." The Department accepted its proposals as a reflection of community desires.

The Fund and WPIC and Brazier were not the only community organizations in Woodlawn. They, however, were the most organized. Woodlawn East

Community and Neighbors (WECAN) was a staunch advocate for low-income residents in Woodlawn. It was not represented on the board of directors of the Fund, and it was generally opposed to much of the redevelopment that targeted attracting upper middle-class residents to the area, labelling it gentrification.

The executive director of WECAN, Mattie Butler, was leery of City Hall and its activities. WECAN was outside of the community planning process until community-wide meetings were held. As a result, WECANs input into the overall redevelopment planning process was negligible. Butler noted that her organization has had no influence on City Hall. She reflected, "we do not have a phone with a line right to the mayor's office. Brazier does." If there is influence she continued, "we have microscopic influence, like throwing a pebble into the sea and the effects ripple outward." Butler suggested that their influence has been concerning bringing attention to the need for creating city-wide affordable housing, helping secure empowerment zone status for the neighborhood status, and completing research on empty land in the area.

In 1991 WPIC approached the Chicago Department of Housing (DOH). (At that time a number of current Department functions were housed in the DOH.) The DOH was impressed with the organization of the plans and alternately enthusiastic with the prospect of pursuing the plans. Feasibility and land-use studies were completed by the Department in 1992 to augment the somewhat general WPIC plans.

According to a Department official, Brazier's relationship with Chicago Mayor Richard M. Daley translated into influence on the planning process. According to the Department official, the reason why the city was supportive of WPICs plan was because it contained a "well-formed idea" and it was comprehensive. In addition, WPIC completed much of the necessary planning work by having consultants and architects develop the plans in advance.

The Department official recounted, "there were many compelling reasons to do (redevelopment), it was not only Brazier's influence, but it was logical." Redevelopment also reflected the mayor's redevelopment policy that favored mixed income housing. Brazier promoted market-rate housing construction and rehabilitation of multifamily units for affordable housing.

6.5.2   The Kenwood Process

Kenwood had been the subject of redevelopment plans from the city of Chicago since the 1960s. However, according to Bob Lucas, executive director of the

Kenwood Oakland Community Organization (KOCO), "none ever material-ized because of lack of political will and community support, more so from lack of political will."

In 1989, the City clearly articulated its interest in redeveloping the area: "given the growing outside interest in the neighborhood, it became important that the community and the City, which owns a large percentage of the vacant land in the area, get out in front of events to ensure that any future development in North Kenwood-Oakland is responsive to the needs and goals of local resi-dents (North Kenwood Oakland Neighborhood Planning Process Community Planning Committee Report 1989, p.5) [16]." The outside interest refers to a private developer who conceived a concept plan in 1987 for developing part of the neighborhood that shouldered the lake. (The developer was said to have noticed that the area was the only undeveloped lake front property in the city while flying over Chicago and wanted to build upscale housing near the lake). In addition, the Department also had part of the neighborhood designated as a blighted and vacant project area in order to "dampen private speculation as well as to protect the City's substantial investment in housing rehabilitation in the area" (North Kenwood Oakland Neighborhood Planning Process Community Planning Committee Report 1989, p.5). [16].

Doug Gills, a member of KOCOs board of directors, echoed the sentiments of WECANs Mattie Butler who noted the influence of the Fund compared to KOCO. Gills related that the leader of the Fund, Victor Knight, "could go in (the Mayor's office through) the Mayor's elevator when KOCO had to go through the front door.... Knight and Brazier designed the Woodlawn Redevelopment plan." Indicating the influence of the Fund, Lucas of KOCO lamented, "(KOCOs) proposals may take a year (to be reviewed by the Department), if Brazier does it, it will be done in 30 days."

In Kenwood, community efforts garnered a Conservation Area designation in 1992. The designation gives the City eminent domain powers to acquire pri-vately held vacant land that could be used to encourage the development of new housing. Importantly, the designation mandates the creation of a Conservation Community Council (CCC) which assists the Department in the development of a neighborhood conservation plan and must approve land-use proposals in the conservation area. Illinois State Legislation grants the CCC only advisory powers; however, in practice, the CCC approves or rejects land-use plans on city owned land. Not by de jure action but by de facto action, the CCCs decisions are accepted by the Department and the mayor as fiat.

In Kenwood, the CCC is the community organization that has the most influence on land-use decision making in Kenwood; however, it was not instrumental in making land-use decisions that were intended to benefit the low-income. It favored development of market rate housing and resisted any subsidized housing. Only community service oriented organizations similar to KOCO were able to influence the creation of housing for the low-income by creating the housing itself as a developer. The CCC, unlike the Fund operating in Woodlawn, was barred by state law from acting as a developer. Although land use that is intended to benefit the low-income is included in the Conservation Plan, the CCC favored the development of market-rate, unsubsidized housing in Kenwood.

6.6   FINISHING TOUCHES: OUTCOMES OF THE PROCESS

Listed on Table 1 are the units of housing produced by community organizations. Low-income housing units composed 22 percent of the total units produced among community organizations. Affordable/moderate income refers to family incomes that are 80 percent or less than PMSA median income. Affordable/moderate income families would therefore have incomes at or below $33,396 and have rents or mortgages no more than $918. Market rate housing refers to housing that approximates the average of housing purchase or rental prices across the PMSA for similarly constructed structures and locations. Included in the table are community nonprofit organizations that produced units of housing during the study period.

TABLE 6.1   Total units produced by community organizations in Woodlawn and Kenwood by Category from 1990 to 1997*.

Community organization	Low income[a]	Affordable/ moderate[b]	Market rate[c]	Total units
		Woodlawn		
WPIC		287	80	367
The fund		122		122
WECAN	52	24	17	93
Covenant development	1		16	17
TOTAL	53	433	114	600
PERCENT	9	72	19	
		Kenwood		
KOCO/KODC	117	233		350

TABLE 6.1 *(Continued)*

Community organization	Low income[a]	Affordable/ moderate[b]	Market rate[c]	Total units
TIA	78	79		157
Ariel foundation	2			2
Urban league			25	25
TOTAL	*197*	*312*	*25*	*534*
PERCENT	*37*	*58*	*5*	
	Neighborhood totals			
TOTAL	250	745	139	1,134
PERCENT	22	65	12	

Notes. [a]Low income refers to family incomes that are 80 percent or less than the Primary Metropolitan Statistical Area (PMSA) median income.

[b]Affordable/moderate income refers to family incomes that are 80 percent or less than PMSA median income.

[c]Market rate housing refers to housing that approximates average of housing purchase or rental prices across the PMSA for similarly constructed and located structures.

*Figures provided by community organizations active in neighborhood areas.

6.6.1   Covenant Community Development

Covenant rehabbed many units that ranged from one single family home to multiunit buildings. It was financed with HUD Hope III money and the Chicago Abandoned Property Program (CAPP) in which the city transferred abandoned properties to parties who will renovate the property. It has also renovated a 10-unit building, and a 6-unit coop, and a rent assisted unit as affordable housing. Covenant as land owner in the development also partnered with WPIC and a private developer in a program that offers homes beginning at $200,000.

6.6.2   The Fund

The Fund partnered with a private developer to rehab two vacant buildings into 102 rental apartments from 1 to 3 bedroom units that were financed through a private bank, the DOH, Federal Home Loan Bank, Illinois Housing Development Authority (IHDA), and an Empowerment Zone grant. It has also completed new home development tagged as affordable with homes ranging from $99,000 to 150,000. The developments were funded through a New Homes for Chicago program where developers received a $30,000 subsidy per home from the City to keep housing affordable.

6.6.3   WPIC

WPIC partnered with a private developer and rehabbed six vacant buildings into 117 units of Low-Income Housing Tax Credit housing with "very low rents" under the Affordable Rents for Chicago (ARC) program. It is financed through the DOH, HOME program, The Federal Home Loan Bank, and the Chicago Equity Fund. The $12 million renovation project was a joint venture with Neighborhood Reinvestment Resources, the Chicago Low-Income Housing Trust Fund, Chicago Equity Fund, private lenders, and the use of Low-Income Tax Credits. More than 70 percent of the initial residents in the buildings lived in Woodlawn before moving into the buildings. Nearly 65 percent of the residents had income below $18,000.

In a similar joint venture as above, WPIC rehabbed seven buildings into 84 total units that included studio and one-bedroom apartments. Renovations were funded by a private bank, the National Equity Fund, and the Federal Home Loan Bank. Another development of 86 rental units was orchestrated by using Low-Income Housing Tax Credits coupled with financing from IHDA and the Chicago Equity Fund.

6.6.4   KOCO/KODC

The majority of KOCOs developments were for the low and moderate income. The development arm of KOCO is the Kenwood Oakland Development Corporation (KODC). Of the 350 units KODC developed, 70 were new construction. A project which produced 117 units for low-income families in 6 buildings was financed by DOH $2.5 million in CDBG funding to City Lands Community Investment Corp, The Illinois Affordable Housing Trust Fund, and the National Equity Fund. Annual income of the tenants ranged between $5,000 and $10,000.

6.6.5   Chicago Urban League

The Chicago Urban League was focused on providing new affordable homes to families in Kenwood. The homes started at $94,500. It was financed through a private bank and DOH New Homes for Chicago housing subsidies.

6.6.6   Travelers and Immigrants Aid (TIA)

TIA, a nonprofit social service agency based outside of Kenwood that serves the entire city of Chicago, partnered with a private developer in a 157-unit rehab of the former Sutherland Hotel. The units for the low and moderate income were financed without direct federal funds as TIA placed $800,000 of equity into the project. Oakwood development Co., a private rehab and construction company that orchestrated the project, added $300,000. The Chicago Equity Fund, which places corporate money to low-income housing development projects by syndicating low-income tax credit benefits, provided $1.2 million. IHDA loaned $500,000.

6.6.7   Ariel Foundation

The Ariel Foundation, a not for profit philanthropic organization, built two affordable single family homes priced to sell at $74,000 to families earning between $24,000 and $36,000 a year. The buyers were to be selected by a committee of local residents, religious leaders, and representatives from the foundation [17]. The housing development incorporated youth who assisted with construction as part of a Careers for Youth program. The foundation collaborated with Uptown Habitat for Humanity as the development's construction company and Careers for Youth as the general contractor which incorporated high school students as laborers.

6.7   RESEARCH FINDINGS

Overall, land-use decision making is subject to the political will of the mayor. At the neighborhood level, land-use decisions are based on the interaction of several factors that include aldermanic preferences, the availability of land, secured funding for development on the land site, and community approval.

6.7.1   Independence versus Influence

The Department's direct influence over land use is minimal as it is limited by mayoral directives and is moderated by developer interests and community participation. The Department responds to directives; it does not set policy. An example of these directives is in the creation of the Fund. In Woodlawn, Rev. Brazier had the support of Mayor Daley who shared a desire for mixed-income redevelopment and Valerie Jarrett, the then commissioner of the Department, who was an appointed official, not a trained, professional planner. Others at the Department, nonmayoral appointees and professional planners, were concerned about possible conflict of interests issues as the Fund wanted to be a "clearinghouse" for redevelopment plans as well as being a developer. According to an official at the Department, at one time, it was Brazier's desire for the Fund to be the only developer in the area. Ultimately the Fund was granted the ability to award development grants as well as to act as a developer.

Daley and WPIC shared desires of creating mixed-income redevelopment that focused on the building or rehabbing of single family homes. In November 1993, Daley announced at a ground breaking for 28-market-rate single family homes, "it has to be mixed. You cannot go back to the old way of rich and poor [18]."

6.7.2   General Plans and the Role of the CCC

The Woodlawn Redevelopment Plan and the North Kenwood Oakland Conservation plans suggested that efforts should be made to utilize land for the provision of housing for the low-income. Neither plan contained specific provisions for the creation of low-income housing. A percentage of development intended for low-income or specific numbers of units to be built were not included. The plans, which also lacked specific numbers or percentages for the development of market rate or other housing, were broad in scope. Though no specific provisions were made, the plans did, however, create a framework for redevelopment that included the use of land intended to benefit low-income residents. The plans suggested mixed-income development. The units to be developed for the low-income were to result from the efforts of nonprofit community organizations and to a limited extent from profit developers.

In Kenwood, land-use proposals which were to bring new housing that was priced $150,000 or above met little, if any, resistance from CCC members, as long as developers and contractors met minority and women participation rates. Affordable housing plans received resistance to the extent that some CCC members were reluctant to approve land-use plans that allowed for the building of "cheap" housing. What received substantial CCC and community resistance were scatter site housing plans. Proposed in late 1995, a plan to locate 241 scatter-site HUD/CHA subsidized units in Kenwood was opposed by the CCC. The CCC was not dedicated to making decisions to utilize land for the low-income. Members of the CCC were reluctant to approve 26 units of low-income housing, down from the original plan of 241 scatter-site units. In effect, the CCC was used as a tool to protect and increase property values of property owners.

6.7.3   Community Developers

The research revealed that only those community organizations that act in the role of developer have influence on the decision making process that results in land use for the low-income. As developers, community organizations must have specific plans for land use and secured financing to develop the plans. Having a land-use plan without financing had no impact on outcomes.

In the two neighborhood study areas, Woodlawn and Kenwood, the City did not respond to low-income housing advocates who proposed that housing for the low-income should be built or respond to opposition during community meetings to plans to build upscale housing by deciding to build low-income housing. This type of community pressure may lead to inclusion and participation in the planning process but does not result in the production of low-income units.

The hypothesis of the study that suggested that land-use decisions intended to benefit the low-income would be a direct result of political conflict turned out to be erroneous. Decisions to utilize land for the intended benefit of the low-income resulted not from the degree of conflict but from the degree of cooperation between communities and political actors. The research here revealed that community groups exert influence only through participation as nonprofit developers and operate much like for profit developers in partnership with the Department.

Land-use decision making followed the standard bureaucratic process of the Department: developers present plans to the Department, the Department works with the developers to refine the plans, and with financing, structures are built.

Conflict with political leaders may lead to participation in planning, but developers build housing. Community pressure produced redevelopment areas and conservation areas, not low-income housing. This finding gives rise to an alternative hypothesis that given a highly organized and mobilized protest movement, community pressure may garner more than participation [19].

6.8   INTERPRETING THE PIECE: SUMMARY AND IMPLICATIONS

Mayoral support and Departmental support for redevelopment plans were relatively easy concessions to community demands. Redevelopment would benefit the city by creating revenue from property taxes on formerly city-held land that was in the hands of developers and homeowners. Redevelopment plans helped to cement political support for Daley in two Black neighborhoods. The support would extend beyond the boundaries of Woodlawn and Kenwood as the neighborhoods mirrored other low-income areas attempting redevelopment initiatives. For the mayor, redeveloped areas would mean improved relations between neighborhoods and City Hall and better relations with Black voters. In addition, supporting redevelopment did not conflict with the mayor's preference for development that focused on attracting and retaining the middle class.

6.8.1   Middle Class Preferences

Woodlawn and Kenwood revealed that community participation reflected a middle-class mien. Both communities, though verbally opposed to displacement of existing residents, resisted developing low-income housing. The question for many of the middle-income was not how to house the low-income, but how not to house them. The middle-income wished to protect their property values and viewed low-income housing as negatively affecting their property investments. Leadership and participation in planning processes also tilt toward the middle-income whose job schedules, flexibility, child care options, and communication styles facilitate inclusion. Those who have lower income tend to be hourly workers who do not have flexible schedules.

6.8.2   A Final Note on Community Influence

As noted, the research revealed that Departmental influence on land-use decisions is minimal and only those community organizations that act in the role of developer have influence on the decision making process that results in land use for the low-income. Participation in the urban planning process does not equal low-income housing outcomes. However, the study revealed that, in Kenwood, where community-wide participation was involved, more low-income housing units were produced than in Woodlawn where community-wide participation in planning was negligible. A KOCO official remarked that Kenwood's process stands in stark contrast to that which occurred in Woodlawn. The development plans were discussed in one community meeting and "the public process came in the form of: "here is the development plan and what color would you like the cover to be?""

REFERENCES

1. P. Peterson, City Limits, The University of Chicago Press, Chicago, Ill, USA, 1981.
2. C. Stone, Regime Politics: Governing Atlanta 1946–1988, University Press of Kansas, Lawrence, Kan, USA, 1989.
3. W. Wiewel and P. Nyden, Eds., Challenging Uneven Development: An Urban Agenda for the 1990's, Rutgers University Press, New Brunswick, NJ, USA, 1991.
4. R. R. Mayer, Social Planning and Social Change, Prentice Hall, Englewood Cliffs, NJ, USA, 1972.
5. P. Clavel, The Progressive City, Rutgers University Press, New Brunswick, NJ, USA, 1986.
6. K. Wong, City Choices: Education and Housing, State University of New York, Albany, NY, USA, 1990.
7. T. N. Clark, Ed., Urban Innovation: Creative Strategies for Turbulent Times, Sage Publications, Thousad Oaks, Calif, USA, 1994.
8. B. London and J. J. Palen, Gentrification, Displacement, and Neighborhood Revitalization, State University of New York Press, Albany, NY, USA, 1984.
9. K. Newman and E. K. Wyly, "The right to stay put, revisited: gentrification and resistance to displacement in New York City," Urban Studies, vol. 43, no. 1, pp. 23–57, 2006. View at Publisher · View at Google Scholar · View at Scopus.
10. G. King, R. O. Keohane, and S. Verba, Designing Social Inquiry: Scientific inference in Qualitative Research, Princeton University Press, Princeton, NJ, USA, 1994.
11. W. Sites, "God from the machine? Urban movements meet machine politics in neoliberal Chicago," Environment & Planning, vol. 44, no. 11, pp. 2574–2590, 2012. View at Google Scholar.

12. Melaniphy and Associate, Chicago Comprehensive Needs Analysis, Melaniphy and Associates, 1982.
13. W. J. Wilson, The Truly Disadvantaged: The Inner City, the Underclass, University of Chicago Press, Chicago, Ill, USA, 1987.
14. N. Lemann, "The myth of community development," The New York Times Magazine, 1994. View at Google Scholar
15. E. J. Kaiser, D. Godschalk, and F.S. Chapin, Urban Land Use Planning, University of Illinois Press, Chicago, Ill, USA, 4th edition, 1995.
16. North Kenwood Oakland Neighborhood Planning Process Community Planning Committee Report, Department of Planning, City of Chicago, 1989.
17. F.F. Rabinovitz, City Politics and Planning, Atherton Press, New York, NY, USA, 1969.
18. E. Michaeli, Daley Announces Rebirth of Woodlawn Area, The Chicago Defender, 1993.
19. W. Sites, "Public action: New York City Policy and the gentrification of the lower East side," in From Urban Village to East Village: The Battle for New York's Lower East Side, J. Abu-Lughod, Ed., Blackwell, Oxford, UK, 1994.

CHAPTER 7

A Structured Decision Approach for Integrating and Analyzing Community Perspectives in Re-Use Planning of Vacant Properties in Cleveland, Ohio

Scott Jacobs, Brian Dyson, William D. Shuster, and Tom Stockton

7.1   INTRODUCTION

The US foreclosure crisis and legacy blight in both urban and suburban areas have led communities to demolish structures that are unsightly, are perceived as a haven for criminal activity, or pose other safety risks in the communities where they exist. Analysis of the typical reasons for abandonment by White (1986) and O'Flaherty (1993) indicated that property taxes, timing of foreclosure, maintenance and condition of premises, and regional development patterns all contribute to abandonment and perhaps over-zealous use of demolition. Increased demolition has left tens of thousands of vacant lots across landscapes in cities like Detroit, MI and Cleveland, OH (Goodman 2005). This has brought with it a widespread change in the structure of urban neighborhoods, which on a block-by-block basis can often have a higher proportion of vacant than occupied housing. Typically, there is no particular coordination of demolition activities in cities in the US. Demolition is generally arbitrated on economic factors alone, environmental and social or cultural factors that relate to the potential for redevelopment are rarely considered (Bell and Kelso 1986; Cunningham 2006). In a study of demolition permit applications in Chicago, Dye and McMillen (2007) found that smaller, older homes that were near public transportation and traditional neighborhood centers were disproportionately selected for demolition, which may work against sustaining or developing vigorous neighborhoods in the future. The literature is consistent in recommending careful analysis of the social, economic, and environmental costs or benefits (e.g., improved public safety, irreparable structural deficiencies, reduction of impervious surfaces and stormwater runoff) of demolishing buildings (O'Flaherty 1993; Dye and McMillen 2007; Power 2008; Bullen and Love 2010). By restoring or refurbishing through adaptive reuse (Bullen and Love 2010), buildings can offer multiple benefits after thoughtful investments and updates are made. For engaging in this careful analysis, the literature is also consistent in recommending development of Decision Support Tools (DSTs), specifically those that enable or widen stakeholder engagement in decision processes at the community level.

Regardless of public policy and urban planning initiatives that include demolition as a part of an agenda for urban renewal, there are increasing amounts of neighborhood residential areas converted to vacant land without a well-defined vision for its reuse. While removal of blighted properties may satisfy some objectives toward urban renewal, uncoordinated demolition may have an overall negative impact on the fundamental objectives that urban renewal is seeking to achieve. As part of a response to this situation, several U.S. municipalities and counties have established land reutilization corporations, commonly referred to as land banks. Some examples of U.S. cities with established land banks (other than Cleveland) include Indianapolis, Louisville, Milwaukee, Philadelphia, Dallas, and San Diego. These land banks are intended to acquire foreclosed or vacant properties and clear titles, to consolidate or aggregate properties, and to maximize the potential reuse or redevelopment of these resources. Typically, these land banks can coordinate with neighborhood groups and provide an administrative process to clear the land titles, generally provide much-needed accounting for exactly where and how much vacant land there is, and finally, make this portfolio of land available to interested buyers that may optimize the value of vacant land (Cunningham 2006). The nascent effort to catalogue and market vacant land resources is concurrent with an emerging movement towards leveraging available land towards environmental restoration and management imperatives. Some strategies that may leverage vacant land portfolios include using vacant lots that are retrofit with or used as part of green infrastructure (GI). GI is defined as infrastructure that uses natural hydrologic features to manage water as a sink for stormwater runoff and to provide environmental and community benefits (USEPA 2010). Strategies for reuse of vacant land that support these greater goals of sustainable development inherently include multiple, complex, compounding and confounding decisions and this process of making decisions should include both decision maker and stakeholder participation. Developing or enhancing GI includes options that must be weighed in context with the entire suite of potential alternatives including the simplest one, which is to do nothing other than to preserve the resource of vacant land until a decision can be made. Under this "holding strategy," these properties are held for future use and economic growth, but not without cost. Processes for considering and understanding complex decisions such as this are needed and an experimental process to support such an effort was undertaken and is described herein.

Environmental and economic problems in the city of Cleveland are typical of the types of problems facing many legacy cities around the US. These problems include deteriorating infrastructure, shrinking urban residential populations, declining industrial and commercial tax bases, poverty, crime, and environmental degradation. In addition to having large amounts of vacant land, Cleveland's assets include the city's geographical location on major transportation routes, strong and vibrant communities, irreplaceable historical buildings, the expansive GI network of Cleveland Metroparks, and plentiful fresh water resources such as Lake Erie and the Cuyahoga River. The city itself, through ownership of vacant properties via the City of Cleveland Land Bank Program, is in the difficult position of considering how best to manage or use this new resource of vacant land. Repurposing vacant land does not constitute a single decision, but many smaller decisions that are arrived at on the basis of multiple, interacting objectives. Each alternate land use or option will include potential advantages and disadvantages, and requires negotiation and cooperation between a large and diverse group of citizens, stakeholders, and decision makers in order to arrive at the best use (or reuse) of limited resources. For urban areas in decline that wish to "re-imagine" whole neighborhoods, development of thorough and transparent processes for making decisions related to the use or reuse of land resources are becoming critically important. Frequently, there is an abundance of data and information known about vacant properties or land targeted for reuse, yet there are few structured approaches to guide reuse of vacant lots. For a logical and transparent decision process, a system is needed that incorporates the information and data known about the land with community involvement to select sustainable alternatives. The structured decision making process (SDM) (Gregory et al. 2012) espouses following prescribed decision steps combined with analytical tools for integrating factual or technical information with stakeholder perspectives. SDM facilitates practical adaptation of available information within a structure to instill methodological rigor and promote credibility. The aims of this paper are twofold: 1) present the USEPA developed beta-version of the decision analysis tool Maximizing Utility for the Reuse of Land (MURL; www.clemurl.org), and 2) evaluate current Cleveland land use information in light of the SDM process, and suggest how it can be tailored to SDM for land reuse planning for a single neighborhood in Cleveland, specifically Slavic Village.

7.2   APPROACH AND TOOL DEVELOPMENT

7.2.1   Decision-Making Approach

An SDM approach is beneficial for addressing multi-stakeholder, multi-objective problems in order to promote clarity, transparency, rigor, and inclusiveness. The following steps describe the main features in the approach (Gregory et al. 2006):

- Define the decision context—Determine and map the economic, environmental, and social drivers, governance structures, regulatory considerations, and stakeholder concerns relevant to the decision problem;
- Identify objectives and preferences—Clarify the values and success measures important and meaningful to decision-makers, and prioritize those with more importance;
- Identify alternatives—Create a range of alternatives intended to meet objectives, reflective of differing perspectives;
- Evaluate consequences—Rank alternatives through quantitative modeling of the problem;
- Conduct sensitivity and value of information analysis—Identify model components most sensitive to new information and determine the value of new information for better decision-making.

Each decision problem is different and the level of effort for each step should be adapted to the needs of the problem, available time, and resources.

The general approach is iterative, given that as understanding of the problem improves changes to prior steps may be required. In practice, it is unusual for one pass through the process to be sufficient for decision-making (Gregory et al. 2012).

Consistent with SDM, MURL incorporates a value-focused thinking approach to decision-making (Keeney 1992). A value-focused approach first asks what stakeholders value and then finds decision alternatives that retain or enhance those values. This is opposed to an alternatives-focused approach that first asks what decision alternatives are available. Thus, a key component of MURL is development of fundamental (ends) and means objectives that reflect stakeholder values. An objective statement includes a decision context, an object, and a direction of preference. The terms "maximize" and "minimize" are often used to indicate direction of preference (Tables 1-2).

TABLE 7.1   Results of survey preference elicitation for fundamental objectives based on an importance ranking of each fundamental objective from lowest (1) to highest (9). The Importance column presents the fraction of the maximum possible priority score such that if all responses for an objective were 9, the Importance would be 1. The Weight column scales Importance to sum to 1. Weight is (w_k) used in Equation 1.

Objective	1	2	3	4	5	6	7	8	9	Responses	Importance	Weight (w_k)
Maximize Social and Cultural Opportunities	1	0	1	0	1	1	3	0	0	7	0.57	0.111
Maximize Environmental Safety and Quality	1	2	2	0	1	1	0	0	0	7	0.35	0.068
Maximize Economic Health and Energy Efficiency	2	0	1	0	1	0	0	2	1	7	0.56	0.108
Maximize Educational Opportunities and Facilities	0	1	0	3	1	0	2	0	1	8	0.58	0.113
Maximize Neighborhood Recreation and General Quality of Life	0	0	0	0	1	1	1	1	3	7	0.84	0.163
Maximize Neighborhood Crime Prevention and Safety	0	1	0	2	1	1	0	2	1	8	0.64	0.124
Maximize Transportation Efficiency	1	0	2	1	0	1	1	0	2	8	0.58	0.113
Maximize Preservation of Historic Architecture and Landmarks	0	3	1	0	1	1	0	2	0	8	0.50	0.097
Maximize Sustainability	2	0	0	1	1	2	1	1	0	8	0.53	0.102

TABLE 7.2 Example results of survey-based preference elicitation for means objectives associated with the Maximize Neighborhood Crime Prevention and Safety fundamental objective. Importance is the fraction of the maximum possible importance a sub-objective could be given.

Maximize Neighborhood Crime Prevention and Safety	Importance
Maximize safety standards in local zoning codes	0.69
Maximize safety standards in building codes	0.69
Maximize lighting along public streets	0.77
Maximize areas open to surveillance (i.e. windows, porches) along public streets	0.74
Maximize security patrols in business districts	0.86
Maximize police presence/visibility in residential areas	0.86
Maximize video surveillance	0.71
Maximize education programs for safety precautions	0.63
Maximize education programs for crime deterrence	0.69
Maximize accurate information on crime levels	0.71

MURL embodies a process that establishes a methodology and a platform for considering all options toward these objectives and allows participants in this process to weigh and consider those options. Fundamental objectives are refined to a point that decision criteria can be established that provide a measure of how well the objective is being met. Means objectives are actions intended to affect the decision criteria connecting "means" to a fundamental objective. Objectives were developed as per Keeney (1992), that were intended to be:

- Complete, so that all of the important consequences of alternatives in a decision context can be adequately described in terms of the set of fundamental objectives;
- Non-redundant, so that the fundamental objectives should not include overlapping concerns;
- • Concise, so that the number of objectives and sub-objectives should be the minimum appropriate for quality analysis;
- Specific, such that each objective should be specific enough so that consequences of concern are clear and criteria can readily be selected or defined; and
- Understandable, so that any interested individual knows what is meant by the objectives.

7.2.2 Tool Development

A requisite model approach was used to develop the decision tool in a way that attempts to contain everything that is essential for solving the issue at hand (Phillips 1982). This approach provides direct and explicit links between what stakeholders prefer and value (fundamental objectives), a mechanism for achieving those preferences and values (means objectives), and the metric for measuring how well those preferences and values are being met. MURL provides an evaluation of alternatives in terms of which "best" satisfy the fundamental objectives. The subjectivity of judging "best' is contextual, and in MURL this is computed and ranked with multi-criteria decision analysis (MCDA) methods. MURL specifically employs multi-attribute value theory (MAVT), which relates preference to the decision criteria. MURL can also use multi-attribute utility theory (MAUT) (not shown in this paper), which like MAVT quantifies preference as a function of criteria input, but also allows for uncertainty in the measures of the criteria (Raiffa 1968; Keeney 1992; Morgan and Henrion 1990; Pratt et al. 1995; Clemen 1996; Drummond and McGuire 2001; Brent 2003).

Many of the criteria used in MURL are derived from geospatial data, and have little uncertainty associated with them (e.g., the location of bus stops). Thus, for the pilot project phase of MURL, value functions are used in the absence of uncertainty. In the MURL approach, stakeholder preferences for different alternative outcomes of a particular decision criterion are represented through a value function. The value function translates the criterion from its original scale (e.g., distance from bus stop in meters) to a common 0-1 value scale (e.g., if a bus stop is a few meters from a parcel than the value might be 1 while if the bus stop is greater than 1,000 meters from the parcel the value maybe 0) placing all criteria on the same scale and therefore making them directly comparable. Decision alternatives are compared through a MAVT-based score of the alternative impacts on the decision criteria. The MURL Score for each alternative is calculated as

$$score_j = \sum_{k=1}^{K} Wk \sum_{i=1}^{Ik} v_i(x_{i,j}) / I_k \qquad (1)$$

where:
j is a policy alternative or decision option
k is a fundamental objective
K is the number of fundamental objectives
w_k is the preference weighting of a fundamental objective

i is a criterion

I_k is the number of criteria for subobjective k

v_i is the value function for criterion i

$x_{i,j}$ is the decision option j's magnitude impact on criteria i

The MURL Score is calculated (Equation 1) by predicting the change in the criteria $(x_{i,j})$ produced by the decision alternatives, normalizing the predicted decision criteria by their value function (v_i), and calculating the sum weighted by the stakeholder objective preference (w_k). The MURL Scores can therefore be used as a basis for policy and decision making based on decision alternative ranking.

The first step in developing the inputs to calculate MURL scores (Equation 1) is development of an objectives hierarchy based on stakeholder input (Figure 1).

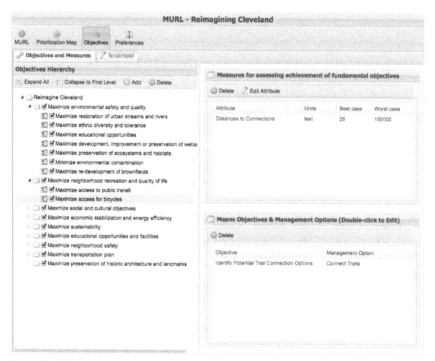

FIGURE 7.1 Screen shot of MURL interface for development of objectives hierarchy. Fundamental objectives are listed to the left. Associated sub-objectives (highlighted) are linked with a means objective (lower right), and a measurable criterion attribute (upper right). Means objectives are the method selected to achieve a fundamental objective and measurable criteria attribute or attributes that quantify the achievement.

The objectives hierarchy tool asks the user initially to develop broad objectives that are then refined to be specific enough that criteria, means objectives, and associated decision options may be specified. MURL requires that criteria and decision options be added or modified only through the objectives hierarchy so that it is clear what specific objective the criteria or decision options address. Given a set of objectives, stakeholder preferences for these objectives can be elicited and translated into weights (wk) that sum to 1.

Determination of objectives preference is accomplished through a technique known as swing weighting. Swing weighting is an elicitation process which uses a series of steps to help the user first rank the decision criteria associated with objectives and then consider the relative importance of each decision criterion as compared to the one immediately preceding it in the overall rankings. There are both simpler and more complex approaches for evaluating stakeholder preferences; swing weighting provides a nice balance between ease of use and theoretical soundness (von Winterfeldt and Edwards 1986). Though the process requires thought and work on the users part, it can help the user resolve or refine their thinking about overall ranking vs. relative importance. Swing weighting asks the user to undertake a two-step process:

1. Decision Criteria ranking
2. Decision Criteria relative preference

The user is asked to pick one objective-linked criterion, which would result in the largest beneficial change (Figure 2).

That criterion is then ranked highest. The process continues, choosing sequentially, resulting in a complete ranking of the criteria. The MURL elicitation process then asks the user to provide a relative preference for one criterion over another starting with the lowest rank criterion and moving to the highest ranked criterion (Figure 3).

The relative preference (Step 2) elicitation approach reduces sensitivity to the overall decision criteria (Step 1) ranking. For example, if the user had difficulty in choosing among the criteria in the Step 1 ranking, a relative importance weight (Step 2) near one can be assigned, giving the two criteria nearly equal weight. The process starts by eliciting relative weights for the lowest and 2nd lowest ranked criterion. The process is then repeated to assign a relative weight of the 3rd lowest-ranked criterion to the 2nd lowest ranked criterion, the 4th to the 3rd, etc., until the highest and 2nd highest ranked criteria are assigned relative weights. Relative Preference scores and Criteria Importance

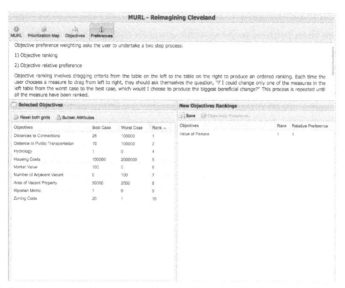

FIGURE 7.2 Swing weights criteria ranking tool. Criteria for fundamental objectives are populated on the left-hand side of the tool. The user then preferentially ranks the objectives on the right-hand side. This is Step 1 of the ranking process.

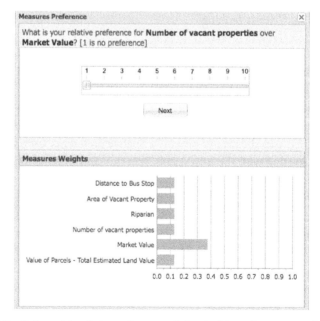

FIGURE 7.3 Swing weight relative preference tool. Step 2 of the ranking process allows the user to better characterize the relative preference between objectives after the initial Step 1 ranking.

Weights (always summing to one) are automatically generated as part of this process. The current status of the weighting scheme is displayed in a bar chart to provide the user with a dynamic visualization of their choices (Figure 3). These objective preference weights are then used in combination with criteria value functions (Figure 4) to calculate a MURL score that can be used to rank decision alternatives.

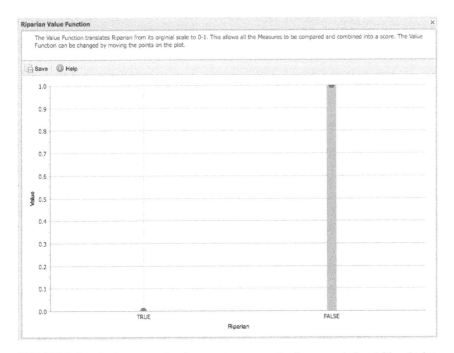

FIGURE 7.4 The Riparian value function is an example of a categorical variable, which is in this case the extent to which a riparian zone is valued by stakeholders. While this function is discrete, value functions can also be continuous (see Figure 5).

The value function for a criterion, $v_i(x_{i,j})$ (Equation 1), specifies a numeric score for each possible level for that criterion that represents the "relative desirability" of each outcome. Figure 5 provides examples of the MURL user interface that allows a user to drag points on the chart to change the shape of a continuous value function for a criterion.

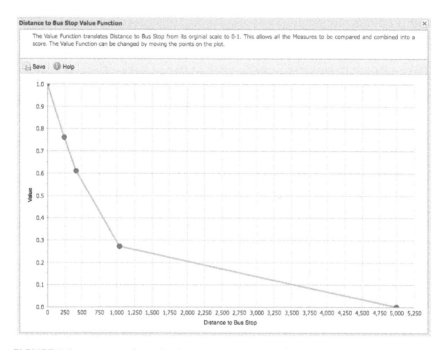

FIGURE 7.5   Distance of parcel to bus stop is an example of a value function for a continuous criterion. As an interface, this function can be altered to investigate the impact on the overall decision and hence is a useful, visual way to communicate preference and priority among stakeholders and decision makers.

The basic decision analysis tools within MURL (objectives hierarchy development tool, value function elicitation tool, and objectives preference elicitation tool) are implemented in a web-based application. The open-source nature of the web application software used in MURL is intended to allow organic growth of a decision support process beyond the original scope and intent of this research once the basic concepts of SDM have been demonstrated. The open source tools used to create MURL include the R statistical programming language (www.r-project.org), PostgreSQL relational database management system (www.postgresql.org), OpenLayers for presentation of GIS information (www.openlayers.org), Geoserver as the GIS backbone (www.geoserver.org), and the ExtJS Javascript library for the web user interface (www.sencha.com).

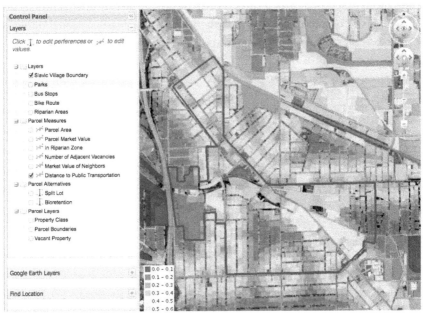

FIGURE 7.6 Example of the MURL mapping tool interface. Displayed is the "Distance to Public Transportation" valuation for each parcel. See Figure 5 for the "Distance to Public Transportation" value function that was applied to the "Distance to Public Transportation" criteria to generate this map.

The MURL mapping tool is a visual interface for geospatial data, criteria data, criteria valuation, and MURL scores for each decision option (Figure 6). The visual interface also facilitates modification of the basis for particular decisions. In particular, the spatial criteria valuation can be updated through modification of the associated value function by clicking on the ✎ icon. Saving changes to the valuation function results in an updating of the underlying criteria value map as well as the associated MURL score map for a decision alternative. The decision criteria that are included in a MURL score for a particular decision alternative can be modified by clicking on the ⵑ icon. This brings up a dialog window that allows the user to add or delete the criteria included in the score (Figure 6). Once such a decision analysis is constructed and implemented, the model is evaluated in one of two ways, which depends on the nature of the inputs. If the inputs are uncertain and specified probabilistically, then a global sensitivity analysis can be performed that identifies the most important factors of the output prioritization. The other option for sensitivity analysis is to change the value of one model input factor at a time. This is similar to performing an iterative "what-if" analysis, and can be helpful when evaluating possible model output for extreme cases. In effect, global sensitivity analysis would identify important inputs, and uncertainty analysis would indicate if there is sufficient confidence in the prioritization, and value of information could be used to determine how much data would adequately reduce uncertainty if the level of confidence is not sufficient in the prioritization.

7.3   ILLUSTRATION OF MURL EAMPLE INTEGRATION AND ANALYSIS OF CLEVELAND LAND REUSE PERSPECTIVES

7.3.1   *Objectives Development and Preference Ranking*

The MURL prototype was applied to the evaluation of alternative land use options for vacant properties in Slavic Village, a neighborhood in the City of Cleveland, OH. A one-day introductory workshop was held in Cleveland, OH to establish a constituency for the project and to ensure that there was interest and investment in model development. With regard to stakeholder input, ideally an explicit and formal process to interact with different stakeholder groups is conducted to elucidate goals and objectives. During the initial workshop with City Departments, it became clear that stakeholders and decision makers in Cleveland had already invested a significant effort to produce a general list of objectives for re-purposing vacant land, had begun to consider alternatives, and had constructed a long-term plan and convened a committee to begin making these difficult decisions.

Rather than starting over and asking these stakeholders and decision makers to work through a process of defining objectives, the research team decided to extract the fundamental objectives, where possible, from the existing plans and reports that the City had already produced. This approach of using approved policy documents for objective development is often termed the gold standard method (Parnell 2007) as opposed to the more time and resource intensive plat-inum standard of formal elicitation interviews with stakeholders. For the pur-pose of this pilot project, objectives were defined using the aforementioned gold standard approach based upon interactions between U.S. EPA and Cleveland city government decision-makers, EcoCity Cleveland (2010), Cleveland Urban Design Collaborative (Cleveland Land Lab 2008), and The City of Cleveland's City Planning Commission's 2020 Citywide Plan (Cleveland 2011b). The overall objective of the decision making process was to maximize improvements to the economic, environmental and social aspects of land use (or reuse) decisions in support of the greater Slavic Village Community Development Corporation (CDC). An initial objective hierarchy was identified that included nine fundamental objectives (Table 1) with a total of 157 sub-objectives (sub-objectives not shown).

Stakeholder preferences for these objectives were elicited and refined through an iterative process with an on-line survey (Preference Step 1) for an initial broad based rapid preference assessment that can be followed by the online MURL swing weighting preference-updating tool (Preference Step 2) when developing alternative preferences for evaluation. The rapid assessment survey allows stakeholder preferences to be collected in a resource efficient manner potentially incorporating a broader spectrum of stakeholders than can be typically gathered in an elicitation workshop. The survey was administered on-line to a group of eight land use managers in local CDCs. The CDCs have the mission of seeking partnerships and providing assistance toward the greater goal of building and maintaining each of Cleveland's neighborhoods. The CDCs that voluntarily participated in this application of MURL were similar in terms of their demographics and the types of challenges encountered. This on-line survey approach was used to quickly rank the set of objectives that could be used to test the MURL methodology and to gain a better understanding of how online surveys should be designed to facilitate ranking of stakeholder objectives efficiently for the first MURL preference step. The limited survey that was done was for proof-of-concept, and the survey results were not intended to support conclusions regarding actual community preferences or objectives beyond their use in development of the beta version of MURL. The survey results form a baseline in MURL against which the user can begin to investigate the impact of values and alternatives on fundamental objectives.

The survey results (Table 1) indicate that the *Neighborhood Recreation and General Quality of Life* objective has the highest priority (Importance score of 0.84 in a normalized ranking from 0-1) for the survey respondents, with the *Neighborhood Crime Prevention and Safety* objective the next highest priority (0.64). *Environmental Safety and Quality* was the lowest priority objective (0.35) for this survey. These preferences may reflect the interest of the participating CDCs in economic development and safety in these urban core neighborhoods, which have experienced a great deal of hardship in the past. The outcome of this limited survey also points out the primacy of economic opportunity as a driver for social and environmental change.

The survey also elicited the importance or relevance of potential means objectives associated with fundamental objectives. Table 2 highlights the means objective preference elicitation for the *Maximize Neighborhood Crime Prevention and Safety* objective from Table 1. The importance of police patrols is evidently seen as important to maintaining overall safe neighborhood environs. Yet, other factors that would directly involve land use decisions or educational approaches to public safety were ranked lower in relative importance.

7.3.2 Scoring and Evaluating Alternatives

The scoring of a decision alternative with MURL (Equation 1) is described at the parcel level with an example involving two objectives, each with a single associated criterion that measures how well the objective is met. Although *Maximize Environmental Safety and Quality* was ranked lowest overall in terms of importance (Table 1), we use this objective with the categorical decision criterion (yes – no) of *In Riparian Zone*, and *Maximize Transportation Efficiency* with the decision criterion of *Distance from Parcel to Public Transportation* (a continuous variable measured in units of length). The *In Riparian Zone* data is derived from an overlay of parcel boundaries and a riparian zone data layer, and each parcel is accordingly assigned a true (in a riparian zone) or false (not in a riparian zone) status. The *Distance from Parcel to Public Transportation* is derived from a distance calculation made between parcel boundaries and a data layer of municipal bus stops. The value functions for *In Riparian Zone* and *Distance from Parcel to Public Transportation* are presented in Figures 4 and 5, respectively.

The importance of *In Riparian Zone* (the criterion for *Environmental Safety and Quality*) is 0.35 and 0.58 for *Distance from Parcel to Public Transportation* (the criterion for *Transportation Efficiency*) (Table 1). Scaling these two values to sum to 1 results in objective weights of 0.38 and 0.62 for *In Riparian Zone* and *Distance from Parcel to Public Transportation*, respectively. Based on this derived information, imagine Parcel A in a riparian zone and 250 meters from a bus stop. The value for *In Riparian Zone* is set to 0 (Figure 4). The implied assumption in setting the affirmative to 0 is that protection of a riparian zone supports the fundamental objective *Maximize Environmental Safety and Quality*. The value for *Distance from Parcel to Public Transportation* is 0.75 (Figure 5). Applying the objective weights produces a score of $(0.0)(0.38) + (0.75)(0.62) = 0.47$. Alternatively, imagine Parcel B, which is not in a riparian zone and 1,000 meters from a bus stop produces a score of $(1.0)(0.38) + (0.29)(0.68) = 0.56$. Under this set of objective preferences, Parcel B would be ranked higher than Parcel A.

Evaluation of alternatives can be demonstrated through examples selected from the "ReImagining Cleveland Vacant Land Reuse Pattern Book" (Kent State 2009), and include:

1. Status quo: take no action on parcel;
2. Split vacant lot among two adjacent owners;
3. Convert parcels adjacent to residences and schools to community gardens;
4. Use parcel as bioretention for managing stormwater;
5. Develop a pocket park as a community garden or a passive green space with seating.

To illustrate how decision alternatives can be evaluated and compared, consider a vacant parcel with 3 occupied parcels surrounding it (Figure 7) with 3 potential decision alternatives (status quo, split-lot, and bioretention), evaluated against four objectives. Table 3 provides the criteria basis for each sub-objective and Figure 7 provides basic parcel characteristics relevant to this example.

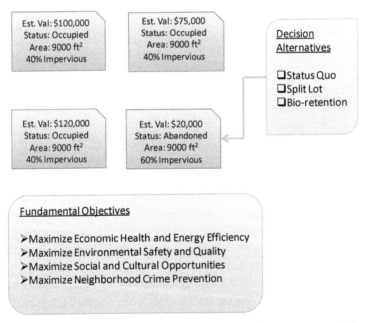

FIGURE 7.7   Illustration of Vacant Parcel Alternatives example. A vacant parcel adjacent to occupied parcels has three alternatives under consideration that must be evaluated using criteria that measure the attainment of fundamental objectives.

TABLE 7.3   Criteria Basis for illustrative example. The example uses four of the nine identified major objectives listed in Table 1. Each criterion is assumed to be linked to a sub-objective of the higher order fundamental objective listed.

Criteria	Assumptions	Units
Maximize Economic Health and Energy Efficiency		
Demolition costs	Abandoned property will be removed for split lot and bioretention options	$10,000 if structure on property ($)
Construction costs	Estimated (Kent State, 2009)	Cost ($)
Property Value Impact	15% reduction in estimated values if not improved	Depressed value of adjacent parcels ($)
	15% increase in value of neighboring parcels if improved	Increased value of adjacent parcels ($)
Maximize Environmental Safety and Quality		
Aesthetics	Stakeholder judgement	High/Medium/Low
Runoff	Equal to vacant lot impervious area for split lot	Reduction in available Run-off surface area (ft²)
	For bio-retention, include impervious areas from adjacent lots	
Maximize Social and Cultural Opportunities		
Parcel ownership	Assigning active ownership to properties is beneficial.	Yes/No
Maximize Neighborhood Crime Prevention and Safety		
Reduce crime	Presence of a vacant structure increases crime.	Yes/No

We assumed that the existence of the abandoned structure depresses the value of neighboring parcels; the split lot option reduces runoff by the area of impervious surface that is removed; the bioretention option redirects flow from the impervious portions of neighboring parcels as well as the target parcel. Given the criteria definitions in Table 3, the impacts of the decision alternatives on each criterion were estimated in Table 4. Table 5 provides the objective weights based on the objective importance from Table 1, normalized for the subset of objectives considered in this example.

The normalized criteria values are calculated by applying value functions (assumed in this example) to each of the measured criteria. This provides the scaled component (i.e., the $v_i(x_{i,j})$) of the MURL score equation (Equation 1). The MURL score is calculated by multiplying the objective weight by the criteria value and then summing for each decision alternative. Comparing the MURL

TABLE 7.4 Decision Alternative Impacts on Criteria $(x_{i,j})$. Results are generated from conditions defined in Table 3.

Objectives & Measures	Status Quo	Split Lot	Bioretention
Maximize Economic Health and Energy Efficiency			
Demolition Costs	0	$10,000	$10,000
Construction Costs	0	$5,000	$29,000
Property Value Impact	-$44,250	$44,250	$44,250
Maximize Environmental Safety and Quality			
Aesthetics	0	0.5	1
Runoff		5400 ft².	16,200 ft²
Maximize Social and Cultural Opportunities			
Ownership	0	1	1
Maximize Neighborhood Crime Prevention and Safety			
Crime	0	1	1

TABLE 7.5 Development of MURL alternative scores. Scores are the sum of the value functions $v_i (x_{i,j})$, weighted by stakeholder preference (w_k). Value functions generate normalized scores from criteria $(x_{i,j})$ with user defined functions (assumed here). See Figures 4 and 5 for examples. Italicized objective weights are re-scaled for the smaller set of objectives in the vacant parcel example and divided by number of subobjective criteria (I_k).

Objectives & Criteria	Objective Weight W_k	Criteria Value Function $v_i (x_{i,j})$		
	From Table 1	Status Quo	Split Lot	Bioretention
Maximize Economic Health and Energy Efficiency	0.108			
Demolition Costs	0.088	0.5	0.2	0.2
Construction Costs	0.008	0.5	0.3	0.05
Property Value Impact	0.088	0.0	1.0	1.0
Maximize Environmental Safety and Quality	0.068			
Aesthetics	0.083	0.0	0.5	1.0
Runoff	0.083	0.0	0.2	1.0
Maximize Social and Cultural Opportunities	0.111			
Ownership	0.270	0.0	1.0	1.0
Maximize Neighborhood Crime Prevention and Safety	0.124			
Crime	0.300	0.0	1.0	1.0
MURL Score		**0.09**	**0.76**	**0.85**

scores indicates that the preferred decision option for this parcel is to convert the parcel to a bioretention basin for managing stormwater (Table 5). The bio-retention decision option appears to dominate the status quo, but the split lot option has a MURL score close enough to warrant further investigation through sensitivity analysis. A sensitivity analysis on the impact of the objective weights and the criteria value functions on the MURL scores could reveal whether the bioretention decision option is a clear choice or whether, for example, the objective weights gleaned from the survey should be updated and refined using the MURL swing weight elicitation tool.

7.4   DISCUSSION

The long-term fundamental goals for redevelopment expressed in the Cleveland Planning Commission's "Connecting Cleveland; 2020 Citywide Plan" are laud-able; however in the immediate term economic considerations and realities tend to greatly outweigh the city's greater ambitions. Therefore, environmen-tal improvement and restoration—which were drivers for the development of MURL—are likely subordinate to making vacant lot space more productive in terms of economic stabilization or improvement. Another assumption that we subjectively impose on this model is that restoration of vacant lots may center on GI. This can take the form of plant-soil systems (e.g., rain gardens) or engineered approaches to realign the local urban hydrologic cycle to emphasize rainfall cap-ture by preventing runoff formation and by providing infiltration opportunities. A lack of familiarity with GI and the services that it can render (stormwater man-agement, green space where there was once none, increased pollinator activity, etc.) may have contributed to a more or less singular focus on economics and safety. In terms of economic interests, the survey respondents recognized the potential to increase the financial value of vacant land by siting a business or residence there, but may not have recognized the intrinsic value in the use of vacant land as a stormwater sink. In the latter case, vacant land becomes part of the regional sewer system with the intent to prevent stormwater from entering combined sewers and treatment plants. This latter arrangement utilizes vacant land as a key ingredient in the management of combined sewer overflows, pro-viding a forum to potentially elevate market value of vacant land for services thus rendered.

   On the matter of safety, perceptions of land use with regard to crime are diver-gent and largely anecdotal. A business or residence in good condition and that provides recognized, real services to the local community is likely to be viewed in a positive light, though building a new business or residential development on vacant land may not be feasible due to overall depressed economies throughout urban core areas. If we apply our normative perspective that GI is a reasonable holding strategy to stabilize the vacant landscape, it is often perceived that tall grasses, trees, and other natural features may provide cover for criminals to hide or conduct illicit trade, among other undesirable social behaviors. A rare field example of this uncertainty in how GI may or may not contribute to safety shows that this is a complex issue that requires further study. The work of Gorham et al. (2009) indicates that increased green space from Houston, TX community gardens (a form of GI) was a potential driver for maximizing the perception of safety, which may influence a community-supported decision making process. Though there were no significant differences in actual numbers of property crimes committed near gardens or in other randomly selected areas, residents of the community garden areas perceived their respective neighborhood areas to be safer due to the presence of community gardens. Studies such as this could be used in targeted educational efforts to connect potential methods for environmental improvement to stated preferences for land use. For the same reasons, the work of Gorham et al. (2009) requires replication in other areas with different demographics to help make clear connections between actual shifts in land use to, for example, GI and the social and economic response that may follow its implementation.

7.5   CONCLUSIONS

The process for developing and applying a rigorous and thorough decision support tool or process to any complex problem begins by defining the decision context and establishing fundamental objectives. Ideally, this process is tailored to the specific site or problem being addressed. MURL represents a generic process that was tailored to repurposing vacant land in Slavic Village. To accomplish tailoring of a generic process to a specific site or problem workshops, meetings, interviews, and literature reviews are conducted. As this process unfolded for MURL, the research team learned that much of the groundwork needed to develop fundamental objectives had already been done and published by decision makers and stakeholders in Cleveland. We realized that the ideal decision support process should place a minimal burden upon the stakeholders and decision makers. Decision-making is hard work, but MURL provides stakeholder value elicitation tools coupled with an underlying rigorous SDM framework that conveniently provides trade-off analysis of decision alternatives. Stakeholder value preference structuring can be achieved by extracting fundamental objectives from existing programs, organizations, and publications and using the resulting objective hierarchy as a starting point for the more traditional approach of elucidating the decision context and objectives from workgroups and interviews. Surveys can be designed and used to further refine fundamental objectives and to develop strategic objectives. This substructure then forms the default conditions for GIS-linked visualization and optimization software to support further understanding and exploration of potential alternative land uses or means objectives. In this way, we are tailoring a tool to support decision making using existing progress and momentum to establish and refine objectives and placing that into a systematic method that allows all stakeholders access to and participation in a process to weigh and consider reuse of the resource that is vacant land in an urban environment.

While our intent was to develop a process to consider or optimize options for the reuse of existing vacant land, this same approach could be expanded further to consider the demolition of vacant or foreclosed properties to combat blight. Power (2008) argues that a focus on renovation coupled with highly selective demolition would be a more sustainable approach than large-scale demolition when a holistic accounting for energy use for each approach is taken into account. The most sustainable solution or path for "re-imagining" a neighborhood in decline may hinge initially on the decision to demolish rather than to renovate or retrofit structures, which is, in and of itself, a complex decision based on many smaller decisions with multiple, interacting objectives, each under a state of uncertainty. Further refinement or development of this decisionmaking process could be done to include the initial decision of whether or not to demolish or leave vacant structures in place as part of "re-imagining" or optimizing land use for a geographical unit at the neighborhood, city, or regional scale. In Cleveland, specifically, two land banks hold vacant lots or lots with structures. The Cleveland Land Bank holds vacant lots and lots with vacant structures greater than 3000 sq. ft, whereas the Cuyahoga County Land Bank holds vacant lots with structures under 3000 sq. ft. When a structure on a lot is demolished, the vacant lot is transferred to the Cleveland Land Bank for disposition. Under this arrangement, the first step in the decision making process will be to determine which land bank will have control of the property and therefore act as decision maker regarding demolition. To consider whether to demolish a structure, stakeholders may need to work with multiple land banks and an expansion of the SDM process described herein could be used.

MURL is scale-independent in theory while in practice the data requirements and stakeholder population grows as the scale grows. At the neighborhood scale, the stakeholder group is narrowly defined compared to a regional scale, but the use of a refined survey approach may help to alleviate this hurdle. Geospatial data also becomes more difficult to collect and manage as the scale increases, but again this hurdle is being lowered over time as federal, state, and local government agencies develop interoperable geospatial data products.

It is important to highlight the fact that MURL is a tool or platform on which stakeholders may consider and compare disparate options; however, it is not meant to result in a "master plan" or to rank alternatives in a static way. As a process, the intent is to inform decision makers as they weigh alternatives – not to dictate the optimal alternative on a site-by-site basis, but rather to compile, compare, contrast, and consider options with the relevant information that is available and with some idea of the uncertainty involved and how that may affect a desired outcome. Information with an unacceptable level of uncertainty can be identified and either omitted from the analysis or highlighted as an area needing further study or refinement. Stakeholders and decision makers can apply this tool individually or in concert to consider options and to investigate, in a defensible process, how objectives may compete or interact and to interpret the results as part of an open conversation held in a visual and intuitive GIS-linked format.

Fundamental to the MURL approach to decision analysis is the iterative learning and decision framing philosophy that occurs as objectives and associated values are elicited. Though a score is calculated that is a valuable guide to ranking decision options, the process of understanding values, designing decision options, evaluating decision options in a manner that is directly and explicitly tied to objectives, and generally thinking hard about the decision problem at hand in a rigorous decision framework is invaluable in moving towards sustainable decisions for the reuse of vacant land.

REFERENCES

1. Bell, E. J., Kelso, D., 1986. The demolition of downtown low-income residential buildings: A discriminant analysis. Socio. Econ. Plan. Sci. 20 (1), 17-23.
2. Brent, R.J., 2003. Cost-Benefit Analysis and Health Care Evaluations. Edward Elgar,
3. Cheltenham.
4. Bullen, P.A., Love, P.E.D., 2010. The rhetoric of adaptive reuse or reality of demolition: Views
5. from the field. Cities 27, 215-224.
6. Clemen, R.T., 1996. Making Hard Decisions. Duxbury, Pacific Grove.

7. Cleveland 2011a. City of Cleveland "Community," Mission Statement. http://www.city.cleveland.oh.us/CityofCleveland/Home/Community (last accessed 6/19/13).

8. Cleveland 2011b. 2020 Citywide Plan, Plan and Implementation. City of Cleveland, City Planning Commission. http://planning.city.cleveland.oh.us/cwp/SummaryImp.php. (last accessed 6/19/13).

9. Cleveland Land Lab, 2008. Re-Imagining a More Sustainable Cleveland: Citywide Strategies for Reuse of Vacant Land. Cleveland Land Lab, Cleveland Urban Design Collaborative, Kent State University, Cleveland.

10. Cunningham, C.R., 2006. House price uncertainty, timing of development, and vacant land prices: Evidence for real options in Seattle. J. Urban Econ. 59, 1-31.

11. Drummond, M., McGuire, A., 2001. Economic Evaluation in Health Care: Merging Theory with Practice. Oxford, Oxford.

12. Dye, R.F., McMillen, D.P., 2007. Teardowns and land values in the Chicago metropolitan area. J. Urban Econ. 61, 45-63.

13. EcoCity Cleveland. 2010. Ecological Design. https://tcfonline.clevelandfoundation.org/catalog/org.shtml?org_id=7875 (last accessed

14. 6/19/13).

15. Goodman, A.C., 2005. Central cities and housing supply: growth and decline in US cities. J. Hous. Econ. 14, 315-335.

16. Gorham, M.R., Waliczek, T.M., Snelgrove, A., and Zajicek, J.M., 2009. The impact of community gardens on numbers of property crimes in urban Houston. Hort Technology, 19(2), 291-296.

17. Gregory, R.S., Failing, L., Ohlson, D., & McDaniels, T.L., 2006. Some pitfalls of an overemphasis on science in environmental risk management decisions. J. Risk Res. 9, 717–735.

18. Gregory, R.S. Failing, L., Harstone, M., Long, G., McDaniels, T.L., and Ohlson, D., 2012. Structured Decision Making: A practical Guide to Environmental Management Choices, Wiley-Blackwell, Oxford,

19. Keeney R.L., 1992. Value-Focused Thinking: A Path to Creative Decisionmaking. Harvard University Press, Cambridge.

20. Kent State. 2009. Re-imagining Cleveland; Vacant Land Reuse Pattern Book, Cleveland Urban Design Collaborative, http://www.cudc.kent.edu/

21. Morgan M.G., Henrion M., 1990. Uncertainty: A Guide to Dealing with Uncertainty in Quantitative Risk and Policy Analysis. Cambridge University Press, Cambridge.

22. O'Flaherty, B., 1993. Abandoned Buildings: A Stochastic Analysis. J. Urban Econ. 34, 43-74.

23. Parnell, G.S., 2007. Chapter 19, Value-Focused Thinking Using Multiple Objective Decision Analysis, Methods for Conducting Military Operational Analysis: Best Practices in Use Throughout the Department of Defense, Military Operations Research Society, Editors Andrew Loerch and Larry Rainey.

24. Phillips L.D., 1982. Requisite decision modeling. Journal of the Operations Research Society 33:303–312.

25. Power, A., 2008. Does demolition or refurbishment of old and inefficient homes help to increase our environmental, social and economic viability? Energ. Policy 36, 4487-4501.

26. Pratt J., Raiffa, H., Schlaifer R., 1995. Introduction to Statistical Decision Theory. MIT Press, Cambridge.

27. Raiffa, H. 1968. Decision Analysis: Introductory Lectures on Choices under Uncertainty. Addison-Wesley, Reading.

28. USEPA, 2009. Lessons Learned in Building Material Reuse and Recycling in Cleveland. EPA-560-F-09-515. October 2009.

29. USEPA, 2010. Green Infrastructure Case Studies: Municipal Policies for Managing Stormwater with Green Infrastructure. EPA-841-F-10-004, August 2010.

30. USEPA. 2010. Sustainable Urban Land Use & Green Infrastructure Workshop, March 2, 2010, Cleveland.

31. USEPA. 2011a. Sustainability. Basic Information. http://www.epa.gov/sustainability/ (last accessed 6/19/13).

32. USEPA. 2011b. Smart Growth http://www.epa.gov/smartgrowth/index.htm (last accessed 6/19/13).

33. von Winterfeld, D., Edwards, W., 1986. Decision Analysis and Behavioral Research. Cambridge University Press. New York.

34. White, M.J., 1986. Property Taxes and Urban Housing Abandonment. J. Urban Econ. 20, 312-330.

PART III
Tools for Community-Based Urban Planning

Development of Future Land Cover Change Scenarios in the Metropolitan Fringe, Oregon, U.S., with Stakeholder Involvement

Robert W. Hoyer and Heejun Chang

8.1 DISASTER RESILIENCE AND DISASTER VULNERABILITY

Human pressures on the environment have their most apparent manifestation in the visible transformation of the Earth's surface. Over the last 50 to 100 hundred years, the most important factor in the change in terrestrial ecosystems has been land cover conversion [1,2], and this trend is likely to continue in the future [3]. Land use/land cover (LULC) maps offer a way to document and quantify these changes [4]. Technological improvements over the last several decades have enhanced LULC maps' ability to observe the outcomes of social and eco-logical processes on the landscape [5]. Projecting LULC patterns into the future can be a useful exercise for evaluating how these processes change and identi-fying potential consequences. Creating a series of possibilities given the avail-able information can provide insights for spatial planning. These possibilities or scenarios provide a useful way to sketch out the future with a level of plausibility "while explicitly incorporating relevant science, societal expectations, and internally consistent assumptions about major drivers, relationships, and constraints" [6]. LULC change scenarios are important, because these can be used to evaluate the potential environmental impacts of decisions or policy shifts [7,8].

In many instances, scenario creation is expert- and/or model-driven, and researchers make the case for their utility to end users [9,10,11,12]. This is problematic in a case with high stakes and high uncertainty, as with land use. Decision makers often prefer their own judgment to model results, highlighting the need for a model to be transparent and simple [13]. Using a participatory approach can partly relieve this issue. Stakeholder and public participation legitimizes the process and justifies the use of the outcomes for planning and decision-making [14]. A key issue in scenario-building methods is the integration of stakeholder-derived qualitative data (typically in the form of a storyline) into models that require quantitative data to produce the final output [15]. There are strategies proposed to formally bridge this divide, like fuzzy cognitive maps [16]. However, in some practical cases, where LULC change is a large component of the final scenarios, an intuitive conceptual approach is used to translate qualitative storylines to quantitative input [17,18,19]. Our study follows a similar approach to these.

The research objective is to develop maps of future LULC scenarios for the study area involving stakeholders. For this research, we define stakeholders as members of organizations with interests in LULC change within the study area [20]. Other researchers created LULC maps in the region for a larger area using a participatory method [21]. This work required several years with numerous iterations that struck a balance in defining assumptions along a gradient of citizen engagement and expert opinion for mapping outcomes to reach satisfactory results. The trend in scenario-building studies is that they are typically a time-intensive process. In an attempt to produce an LULC map relatively quickly, while still retaining the benefits of a participatory approach, we used a simple framework that integrated input from stakeholders with local knowledge into a geographic information system (GIS) modeling process. We took information gathered from a single workshop, as well as a few additional one-on-one conversations and used these as guiding principles for future land cover change. In this respect, our study is more a consultation than an engagement process, but it is a useful method for addressing an important problem in our research program. Although we anticipate more development along the urban-rural fringe in our study area, it is currently unknown where and how much new development will be placed specifically.

8.2 METHODS

8.2.1 Study Area

The Tualatin and Yamhill basins drain a portion of the Willamette Valley's northwest corner and are 1858 and 2000 km2, respectively (Figure 1). The area holds the three broad land typologies of western Oregon—developed lands, agriculture and natural vegetation dominated by upland forests. A significant portion of the Tualatin basin lies in the greater Portland metropolitan area. Washington and Yamhill counties, whose areas approximately correspond to the majority of the study area, have experienced rapid growth since 1980 (Table 1) [22], continuing a legacy of population growth in the Portland metropolitan area over the last century [23]. The city of Hillsboro, a west Portland suburb, has more than tripled in population between 1980 and 2010. Washington County is seeing higher population density increase than the average for all counties of the north Willamette Valley region (Figure 2) [24]. Despite higher density, urban land cover continues to grow in the study area as population increases (Table 2) [25,26].

TABLE 8.1   Populations by decade for two counties and select cities in the study area.

	Pop. 1980	Pop. 1990	Pop. 2000	Pop. 2010	Ann. Ave. Change	Total Change
Country						
Washington	245,860	311,554	445,342	529,710	2.39%	115.4%
Yamhill	55,332	65,551	84,992	99,193	1.90%	79.3%
Large Cities						
Beaverton	31,962	53,307	76,129	89,803	3.39%	181.0%
Hillsboro	27,664	37,598	70,186	91,611	3.94%	223.2%
McMinnville	14,080	17,894	26,499	32,187	2.70%	128.6%
Newberg	10,394	13,086	18,064	22,068	2.46%	112.3%

To understand the drivers of the area's land cover change, a discussion of the policy context is warranted, as institutional factors mediate people's response to economic opportunities [27]. In 1973, the Oregon legislature created an institutional framework for land-use planning applying state-wide goals informed by citizens and implemented through local governments [28]. This regulatory environment was meant to first protect Oregon farmlands and also encouraged livable urban communities with well-planned infrastructure. A primary tool towards this end was an urban growth boundary (UGB) for each incorporated city to reduce urban sprawl and encourage future compact development. These UGBs effectively create a land use dichotomy of urban lands within and the

FIGURE 8.1   Study area, including the current urban growth boundaries around each municipality.

resource lands on the outside. In 2004, the passage of state ballot Measure 37 allowed a compensation claim or waiver of development restrictions for property acquired prior to the enactment of the legislation [29]. Land use planning advocates convinced voters to replace it with Measure 49 in 2007. It provides a more rigorous definition of compensation by limiting claims to three or less new dwellings on a parcel. In 2007, Metro, the Portland regional governing agency, proposed urban and rural reserve areas (URAs and RRAs) surrounding the current UGB to plan for growth in a manner compatible with state land use goals, which include targeting areas for future development that limit impacts on ecological systems [30]. These basins and the Tualatin in particular have also come under a high level of scrutiny for water quality issues [31,32]. Both basins have stream reaches placed on the state's 303(d) list for impaired surface water bodies in accordance with the federal Clean Water Act. Total Maximum Daily Loads (TMDLs) are in place or in development for several water quality indicators [33]. LULC plays a major role in determining water quality indicator values and stream health within both urbanizing and agricultural catchments in the study area [34,35].

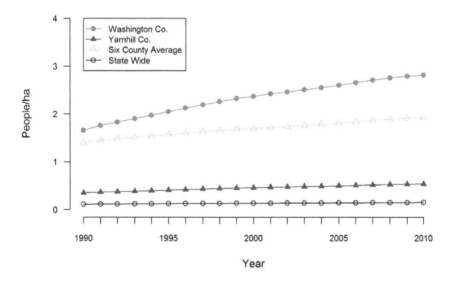

FIGURE 8.2 Annual increase in population density (people/ha) for two counties. The six Oregon counties of the northern Willamette Basin (Clackamas, Columbia, Marion, Multnomah, Washington and Yamhill Counties) are included for reference. Data are from Portland State College of Urban and Public Affairs Population Research Center.

TABLE 8.2   Change in urban land cover classes for two counties and select cities in the study area. Change based on National the Land Cover Dataset (NLCD) 1992–2001 land cover change retrofit product [34] and the 2001 and 2006 NLCD datasets.

	1992 to 2001 Urban Change (ha)	Percent Change	2001 to 2006 Urban Change (ha)	Percent Change
Country				
Washington	1209	4.3%	1073	3.0%
Yamhill	260	20.8%	381	3.0%
Large Cities				
Beaverton	108	2.5%	43	1.0%
Hillsboro	293	6.5%	288	5.8%
McMinnville	34	1.8%	138	7.2%
Newberg	89	8.1%	43	10.9%
Study Area	1669	3.1%	1476	2.7%

8.2.2   Data

We chose the USGS National Land Cover Dataset (NLCD) 2006 as the baseline land covers in our study area for a variety of reasons [26]. First, the dataset contains a manageable amount of classifications, with 15 falling within the study area. Second, at 30-m resolution, it allows a fair degree of spatial differentiation without overwhelming the subsequent modeling efforts. Third, the year 2006 is the most up-to-date product available from the USGS. The socioeconomic calculations for this project started in 2010 to align with U.S. Census estimates. Considering the late decade economic downturn slowing of new development, we assumed the four-year difference in land cover would be small at the landscape scale. Several other datasets were gathered from various state and local agencies and governments [36,37,38,39]. As the focus was on the increase in urban development, most of the data was primarily composed of spatially explicit data pertaining to the Oregon land use regulation framework (Table 3).

TABLE 8.3   Data sources used to create a spatial mask and graded weight map used as inputs to a model, creating future scenario maps in the study area.

Data Type	Description	Source
Urban Growth Boundaries (UGB)	Includes current UGB plus accepted and proposed urban reserve areas (URAs), rural reserves with additional protection and some additional adjacent land in case growth exceeds current reserves.	Metro Regional Land and Information System (RLIS), City of McMinnville Planning Department, City of Newberg Engineering Department
Zoning	Includes all except a few small communities. A statewide layer designating broad classifications (forestry, agriculture and rural residential) was integrated with municipality zoning layers.	RLIS, Mid-Willamette Valley Council of Governments (City of Dayton's zoning estimated from online map)
Measure 49 Claims	632 claims joined to tax lot parcel data to make spatially explicit. Authorized claims collected from three counties making up the vast majority of the study area.	Oregon Department of Land Conservation and Development, State of Oregon Geospatial Enterprise Office, Yamhill County Assessor's Office
High Value Farm Soils	Agriculture soils of U.S. Natural Resource Conservation Service Class I and II (irrigated or non- irrigated)	Oregon Spatial Library
Groundwater Restriction Zones	Critical and restricted groundwater zones could possibly be an impediment to rural residential development. Designated by the Oregon Department of Water Resources where aquifers are identified as depleted or used at an unsustainable rate	Oregon Department of Water Resources
Protected Areas	Lands off-limits to development for a variety of reasons, including federal forest lands, city and state parks, private green spaces and schools.	RLIS, U.S. Fish Wildlife Geospatial Services

8.2.3   Construction of Scenarios with Stakeholder Consultation

Our modeling framework followed a multistage process (Figure 3). Using a previously published approach, we elicited opinions on our modeling process from a small group of stakeholders. The researchers took the view that in their limited role, stakeholders would serve to validate assumptions the researchers made in creating maps, or if they distrusted the results, we could make improvements. Swetnam et al. [18] used a rules-based framework for integrating stakeholder narrative data into a quantitative geographic information system (GIS) modeling method. We produced an initial LULC map projected for the year 2050 using this method with data and rules defined by the researchers. The initial rate of changes in land cover types were estimated through an extrapolation of the differences detected in the NLCD 1992–2001 land cover change retrofit product [25], as well as the 2006 NLCD product.

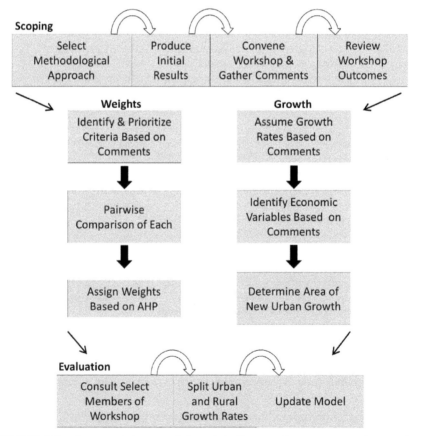

FIGURE 8.3 Conceptual diagram of the stakeholder consultation process informing the development of the modeling approach used to produce future land cover scenarios. AHP, pairwise analytical hierarchy process.

The researchers convened a workshop in June 2012, that lasted several hours. A project partner involved with the current environmental issues in our study area chose four professionals for the consultation. They represented a cross-section of land use interests, including a representative from the Oregon Department of Agriculture, a county planner, an economist with Portland Metro administrative and planning agency and a land use attorney. We presented the project background information and the LULC modeling method and initial maps. Workshop participants initially discussed what they knew about land use in the region. The workshop then evolved into discussion about the future of developed lands in the study area and the factors they considered relevant.

As the discussion progressed, it became evident that the participants' opinions pointed to the state land use regulatory framework being the primary factor deciding where new developed land would be located over the next several decades. It yielded other important points. Any large increases of farmland were unlikely given that almost all suitable lands were already in production. County planners worked hard to maintain rural landscapes, so although conversion of farmlands will occur, they will be too small to fundamentally alter the land cover type present. Based on these participants' inputs, we decided to focus on new urban development in our modeling effort. They quickly agreed that the UGB was the most important factor, followed by the Oregon Department of Water Resources groundwater restriction zones [40], high value farm soils [41], zoning and Measure 49 claims. These factors are readily available or adaptable to spatial datasets (Table 3). In our subsequent scenario maps, participants replaced the criteria chosen by the researcher in the initial map that included a few biophysical factors, like soil type and slope.

For quantifying new urban growth, we relied on the two planner participant comments that made the link between urban growth and the accommodation of new population. They suggested basing the estimated amount of required new urban area on population growth and a few demographic variables. When asked for rates of population growth within the range of plausibility, we received a single volunteered response of roughly 0.5% to 2.5%, which we adapted to 0.6%, 1.5% and 2.0% for the construction of future scenarios. This is supported by the known increases in observed growth over the previous decades (Table 2 and Table 3).

We performed some simple calculations to link population to urban development, based on consultation with planning professionals who participated in the workshop. These included future average household size (2.46) [23], an estimated employment-population (e-p) ratio (0.44) consistent with 2010 population and jobs numbers [42,43], employment per household (1.2) and an estimated density of future jobs and residences. Employment per household was a slight modification of the rule of thumb of one job per household suggested by one. Both planners anticipated a modest increase in job density in the future and smaller residential lot sizes and higher density housing developments in the upcoming decades. The researchers chose an employment density metric slightly higher than a current estimate using the e-p ratio and the NLCD 2006 high development category. This was regarded as plausible, as many employment facilities will continue to be low density, like warehouses. One participant mentioned a current density target for housing (approximately 35 per ha). We chose a somewhat higher figure (42 per ha) to account for the existing urban area absorbing a small portion of the additional needed residences and the additional comment that density targets are likely to increase in the future under political pressure. The additional required land averaged with the current urban land base yielded small to moderate increases in urban densification (Table 4).

To allocate area to NLCD's different developed cover classes, we assumed that the proportional relationship between open, low and medium development would hold from current conditions. The open development class in our study area typically covers urban greenspaces, such as city parks, large lawns and golf courses. Low and medium classes cover the majority of residential areas. A final modification was suggested again by the planners to split the growth between

TABLE 8.4   Summary of the metrics used to calculate area of new urban land cover by the year 2050.

Scenario	Area	Ann. Pop. Growth	Future Jobs per Ha	Total Jobs per Ha	Future Households per Ha	Total Households per Ha
Current (NLCD 2006)	Urban			82.4		8.8
Cureent (NLCD 2006)	Rural			95.3		3.2
Future Low	Urban	0.57%	86.5	83.5	42.0	10.4
Future Low	Rural	0.03%	96.4	95.4	6.2	3.2
Future Medium	Urban	1.43%	86.5	84.3	42.0	13.3
Future Medium	Rural	0.08%	96.4	95.4	6.2	3.2
Future High	Urban	1.90%	86.5	84.7	42.0	14.4
Future High	Rural	0.10%	96.4	95.4	6.2	3.3

* The required area is based on the assumed future jobs and households per ha. The total jobs and households per ha are the density of increase averaged over both current and future urban land cover. Current land cover is based on NLCD 2006.

urbanizing and traditionally rural areas. The sentiment communicated to the researchers was that the regulatory framework would discourage growth in rural areas to the point that it would be very small over the coming decades. At the workshop, participants agreed that very small "cities" are unlikely to expand for cultural, social, economic and infrastructural reasons. As a consequence of this observation, the study area was split into medium to large urban areas and the rest of the landscape. One participant suggested dividing the growth to 95% urban and 5% rural (Table 4). Job densities are higher in rural areas than urban areas, because the job densities were artificially compressed into the small amount of present urban land cover. We summarize the final increases of new developed area from NLCD 2006 to the future scenario in Table 5.

TABLE 8.5   Summary of growth in each developed land cover category in each future scenario expressed as total new hectares and percent increase from the USGS NCLD 2006 dataset*.

Scenario		High Dev.	Medium Dev.	Low Dev.	Open Dev.
Low	Urban	1250 (37%)	603 (5%)	944 (5%)	260 (5%)
	Rural	34 (12%)	5 (1%)	40 (1%)	64 (1%)
Medium	Urban	3046 (91%)	1805 (16%)	2823 (16%)	777 (2%)
	Rural	41 (14%)	13 (2%)	108 (4%)	173 (2%)
High	Urban	4331 (129%)	2665 (24%)	4168 (23%)	1148 (23%)
	Rural	44 (15%)	18 (2%)	146 (2%)	234 (4%)

* Land use categories are based on NLCD 2006.

8.2.4 Mapping

Based on new urban lands dominating the discussion at the workshop, we chose to focus solely on new urban growth. Transitions from existing development types to higher intensity development were not considered because of time constraints and the uncertainty of future densification in the existing developed areas. While we acknowledge that changes within the current urban area will occur, like high density re-development, we assumed they would be small, based on the preservation of the existing residential area structures [44]. Thus, our model only allows new urban land cover to replace agriculture and natural vegetation types, and there are no shifts in the patterns of the remaining agriculture and natural vegetation lands.

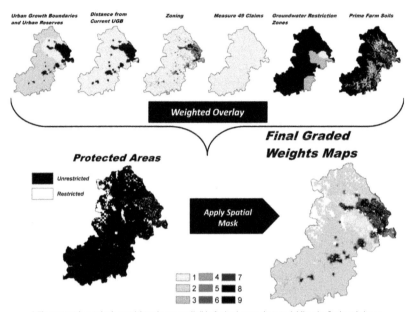

* The protected areas in the spatial mask are not eligible for land cover change, yielding the final graded map
  that is used to guide the assignment of new developed land cover in the study area.

FIGURE 8.4   Maps of the six criteria (urban growth boundaries, distance from current urban growth boundaries, zoning, groundwater restriction zones, high value farm soils and Measure 49 claims) used to construct the graded weights map.

The GIS process, implemented in ArcGIS 10.1 [45], used the combination of a spatial mask based on protected areas and a spatial weight map based on the regulation criteria identified by the stakeholders (Figure 4). The storyline data acquired from the workshop was not sufficiently detailed to address all the assumptions and required parameters. We interpreted the workshop discussion by identifying the criteria that were most emphasized, but also had to use the researchers' own judgment. Weight assignment was performed through a two-stage process. In the first stage, the variables within each criterion were ranked using values from nine, the highest conversion potential, to one, the lowest conversion potential. For example, the UGB criteria layer included the current UGB, URAs, undesignated lands adjacent to the UGB and RRAs. They were ranked nine, eight, five and one, respectively (Figure 4). We included a distance band from the current UGB criterion based on our judgment to preferentially assign new urban map pixels to lands closest to the UGB. Measure 49 claims were incorporated by randomly placing a small group of pixels in a claimed tax lot. This technique is likely overestimating the effect of Measure 49 claims, even though their fraction of the study area is small (~0.15%). Since it was considered an important factor by stakeholders, we did not eliminate it from our analysis. We assumed that the high development land cover class would consume the highest weighted pixels first, followed by the medium class, low and, finally, open. This allowed us to rank the zoning dataset from more intense land use types to least intense.

In the second phase, we calculated weights among the criteria using a pairwise analytical hierarchy process (AHP) [46]. The AHP assesses the importance of each criterion by directly comparing it to all others. For example, we assumed, based on stakeholder discussion and our judgment, that the UGB dataset will be more important than all other criteria types, but some will have more importance to it compared to others. Using the same one to nine value range, our decision was to make the UGB criteria nine times as important as prime farm soils, groundwater restriction zones and Measure 49 claims. It was three times as important as the distance band and twice as important as zoning (Table 6). The weight values were then used to combine all criteria into a single map using a weighted overlay. Finally, we automated the geoprocessing routine using a Python script.

TABLE 8.6   Results of the pairwise analytical hierarchy process (AHP) for each spatial variable incorporated into the final graded map that guides the allocation of new urban land cover grid cells in three scenarios of increased urbanization.

	Urban Growth Boundary (UGB)	UGB Dist.	Zoning	Prime Farm Soils	Groundwater Restriction Zones	Measure 49 Claims	Geometic Mean	Weight
Urban Growth Boundary (UGB)	1	3	2	9	9	9	4.04	44%
UGB Distance	1/3	1	1/2	7	9	5	1.94	21%
Zoning	1/2	2	1	9	7	1	1.99	21%
Prime Farm Soils	1/9	1/7	1/9	1	1/3	1/7	0.21	2%
Goundwater Restriction Zones	1/9	1/9	1/7	3	1	5	0.55	6%
Measure 49 Claims	1/9	1/5	1	7	1/5	1	0.56	6%
Total							9.29	100%

The AHP determines the weight value of each variable to be used in a weighted overlay GIS procedure.

8.3   RESULTS

Our consultation with stakeholders in the study area resulted in a simple storyline. Future urbanization will be placed in the study area where land use regulations allow it to be placed. The urban growth boundary and its planned extensions are the primary factor, but other factors, like zoning, high value farm soils, groundwater restriction zones and Measure 49 claims will also play a role. This qualitative data was transferred to a GIS process through spatial datasets, demarcating where those regulations are enforced. The amount of new urban land cover is harder to address, but a good rule of thumb is to assume population growth will be the main determinant. This is the historical precedent. A suggested range of quantitative values of population growth were used to select low, medium and high urban LULC growth. This basic process produced three maps of urban LULC growth.

The three scenarios maps (Figure 5) showed development increasing along the current urban fringe. In the low scenario, the northern edge of the west side

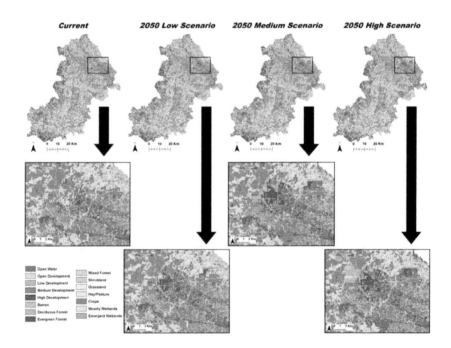

FIGURE 8.5   Maps representing potential future urban land cover change in the study area. The insets represent a portion of the study area showing growth adjacent to the city of Hillsboro, OR.

of the Portland metro area exhibited the most land consumption (Figure 5). The municipality in this area is actively planning for growth as a hub for the technology industry. The spatial regulatory data attracted commercial/industrial or high developed land cover here in all scenarios. Other areas also received growth resembling a "creep" around the edges of the current UGB to accommodate additional housing. This expansion intensified in the medium scenario as commercial/industrial land cover increases more substantially in other areas, including the southern portion of the Portland metro region, as well as the satellite communities in and near the Yamhill basin.

In the high scenario, a large portion of urban reserves were consumed around the western Portland metro area. The southern communities also showed a substantial increase in urban land, and even some of the smaller communities displayed gains. This was illustrated by two of the southern communities beginning to merge (Figure 6). The rural areas and very small communities exhibited very little change over the next forty years considering the very modest growth rates placed on them. This is consistent with the planning goals mentioned in

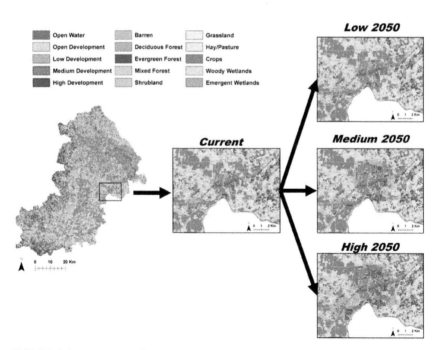

FIGURE 8.6   Inset maps of potential land cover change in the study area focusing on the area encompassing the cities of Newberg (center of inset) and Dundee (southwest quadrant).

the workshop. The simple modeling approach led to unrealistic patterns at the fine scale (Figure 7). Adding refinements to the modeling procedure was not feasible to rectify these discrepancies. The two planning stakeholders reviewed the final maps and confirmed the lack of realistic patterns at this scale. However, at the landscape scale, they agreed that the maps are plausible. The stakeholder

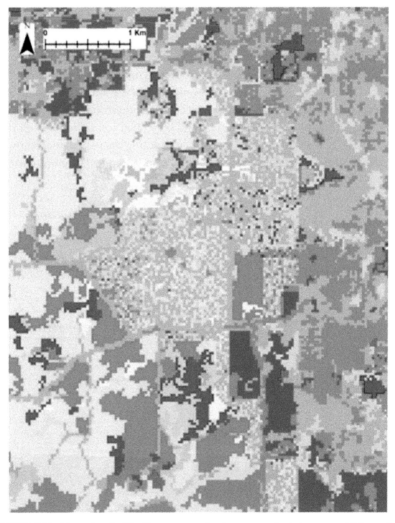

FIGURE 8.7   Example of grid cell raster mosaic phenomena depicting potential future land cover change. Exact arrangement of grid cells is determined randomly where there are more candidate pixels for transformation than are required.

representing agriculture also considered the maps plausible. Considering that this is the scale of analysis that we were most interested in projecting, we felt additional effort to address these issues was not warranted. At a subsequent meeting where the land cover results were presented as a small component of a greater project, we were able to communicate quickly and effectively how scenarios were produced.

8.4   DISCUSSION

This research presents a model of LULC change that provides a plausible answer to our research question of where and how much new urban growth will occur in the urbanizing basins of Oregon. Our expectation was that this simple modeling method could be disseminated easily to stakeholders with at least some familiarity with the geospatial sciences and the characteristics of spatial data. It was our intention to reduce the number of decisions necessary to build and parameterize the model in order to produce maps relatively rapidly. Our final product is not as consistent as those produced through data gathered in a demanding iterative storyline process [15,18,21] or a well-designed role playing game [19,47]. Instead, we gathered qualitative data through an ad hoc workshop and select interviews to develop a simple storyline, which is used elsewhere [48]. The maps can still prove useful as input in further scenario modeling. Considering uncertainty is inherently high in projecting future land conversions, the consistency of the maps is sufficient at the landscape scale. Other modeling approaches, like agent-based systems or cellular automata, may address complexity with more sophistication [49], but also produce a more complex message to explain, leading to additional time investment.

At the workshop where results were presented, members unfamiliar with the process understood it quickly. We aimed for a simple, flexible model that facilitates communication about complex relationships among stakeholders with varying backgrounds [13]. Their comprehension led to questioning an underlying assumption used to produce the final map outputs. Unsurprisingly, it was an assumption based on our judgment: a similar proportion of urban land cover intensities in the future as in the current LULC map. This points out a problem our study shares with others. GIS modeling processes are difficult to fully parameterize with participatory data. This leads to the use of researcher discretion in modifying parameters or to assuming similar values across all scenarios [50]. We attempted to counteract this by relying on the opinion of at least some of the initial stakeholders for the validation of our results. This is similar to the "social validation" approach taken by others [47]. Their agreement that the maps are plausible at the landscape scale lends support to their viability as input in certain applications.

Our consultation process left us with a narrative about increased urban growth and largely assumed other types of land conversions would be minimal to non-existent in our study area. Had various groups with differing perspectives been consulted over the course of a longer process, other factors addressing LULC change in rural areas could have potentially been identified. Even with the urban focus, other variables are likely important. Although deemed not important at the workshop, transportation networks could still be an important driver of land cover change [51]. Groundwater restriction zones, while having real consequences in rural lands, may not be a severe impediment to growth in urban land, since they are fed by surface supplies in our study area [52]. Additionally, stakeholders mentioned infrastructure access variables (e.g., water, sewer, gas lines) as having huge consequences on the location of new development, but data access has proven difficult. Using a few economic/demographic variables as parameters determining the amount of growth is simple and straightforward to communicate. This approach does not account for the dynamic nature of land supply and the spatially variable nature of demand. Our stakeholders pointed out that this is a major factor in urban planning. Our model assumes that as population grows, so will urban development. However, there are concerns that the land use regulations the model is based on will make property values unaffordable to many residents, leading to growth in communities outside of the study area, but still commutable to the employment centers within it. Econometric models based on assumptions of land owner decisions to maximize net returns from land can potentially address some of these issues [53,54].

This analysis hinges on Oregon's land use system being the largest variable in guiding future land conversion in the state, barring any major government policy shifts over the next several decades. Indeed, the analysis already performed by local agencies in defining the current UGB, as well as URAs is what makes the following analysis a practical approach for developing these land cover scenarios and can be thought of as an extension of these efforts [55]. The heavy debt of this work means that a model based on land use regulations will not be generalizable to other regions. What is presented here may indeed rely too much on regulations, considering the land use systems have seen real challenges to their authority [56]. It must then be acknowledged that our scenarios used present conventional assumptions held by stakeholders in the region that aided a relatively quick process of parameterizing our land change model. In this sense, our scenarios are in reality a gradient of outcomes for a single storyline: land use regulations will be the dominant factor in deciding where new urban land will be located. More time and creative thinking would be necessary to develop other alternative possible future realities that aid in planning for the unpredictable [57], but would then challenge the framework for linking qualitative and quantitative data [16]. This issue, in fact, did come up in our workshop, where one member challenged the idea that future urbanization will follow the previous paradigm of what is developed land cover. Stressing technological innovations and social demands, we cannot simply assume that urban lands will impact natural systems through the loss of biota, increased impermeable surfaces, etc., as they did before. Exploring a scenario with these compelling qualitative factors would necessitate a much increased effort to develop a LULC map matching such a vision.

8.5   CONCLUSIONS

The method presented in this paper offers a simplified and transparent approach for producing future land cover scenario maps. Our process contains important stakeholder involvement that was integral to identifying and prioritizing the factors that drove new urban growth in the model. Our main objective was to develop a straightforward process that was easy to communicate. We make several observations describing the degree of success we obtained in our study.

(1) Although simple, our land cover projection requires the use of a quantitative GIS process to actually produce the maps, so it still faces the same issues of other scenario modeling efforts when translating qualitative data to quantitative.

(2) Keeping the stakeholder consultation limited is advantageous, because it allows researchers to model future land cover under manageable time and effort constraints with widely available GIS data. Our land cover scenarios represent a gradient of potential realities based on the same storyline, and researchers still needed to make assumptions and set some of the parameters themselves.

(3) The stakeholder consultation led to place specific analysis (e.g., different growth rates for urban vs. rural areas). The land use regulatory system is unique to Oregon, and the spatial data based on it encapsulates a great deal of external analysis that made our process faster. This highlights the

potential difficulty in generalizing scenario development frameworks that facilitate reproducibility [15].

Ultimately, we acquired an answer to our pressing research question of where will new urban growth be placed and how much of it will there be. We conclude that developed land will consume portions of the metropolitan fringe, and its amount will be determined by how much population will be present in the area by 2050. The case study points out that a relatively simple GIS-based modeling process is possible given available data, but this also leads to sacrificing some complex dynamic processes of land cover change. Therefore, this effort represents an initial step in modeling land cover in our study area, and further modifications and refinements to the participatory framework and to the model itself are warranted.

REFERENCES

1. Millennium Ecosystem Assessment (MA). Ecosystems and Human Well-Being; Island Press: Washington, DC, USA, 2005.
2. Foley, J.A.; DeFries, R.; Asner, G.P.; Barford, C.; Bonan, G.; Carpenter, S.R.; Chapin, F.S.; Coe, M.T.; Daily, G.C.; Gibbs, H.K.; et al. Global consequences of land use. Science 2005, 309, 570–574.
3. Seto, K.C.; Güneralp, B.; Hutyra, L. Global forecasts of urban expansion to 2030 and direct impacts on biodiversity and carbon pools. Proc. Natl. Acad. Sci. USA 2012, 109, 16083–16088.
4. Gulickx, M.M.C.; Verburg, P.H.; Stoorvogel, J.J.; Kok, K.; Veldkamp, A. Mapping landscape services: A case study in a multifunctional rural landscape in the Netherlands. Ecol. Indic. 2013, 24, 273–283.
5. Turner, B.L.; Lambin, E.F.; Reenberg, A. The emergence of land change science for global environmental change and sustainability. Proc. Natl. Acad. Sci. USA 2007, 104, 20666–20671.
6. Thompson, J.R.; Wiek, A.; Swanson, F.J.; Carpenter, S.R.; Hollingsworth, T.; Spies, T.A.; Foster, D.R. Scenario studies as a synthetic and integrative research activity for long-term ecological research. BioScience 2012, 62, 367–376.
7. Bateman, I.J.; Harwood, A.R.; Mace, G.M.; Watson, R.T.; Abson, D.J.; Andrews, B.; Binner, A.; Crowe, A.; Day, B.H.; Dugdale, S.; et al. Bringing ecosystem services into economic decision-making: Land use in the United Kingdom. Science 2013, 341, 45–50.
8. Goldstein, J.H.; Caldarone, G.; Duarte, T.K.; Ennaanay, D.; Hannahs, N.; Mendoza, G.; Polasky, S.; Wolny, S.; Daily, G.C. Integrating ecosystem-service tradeoffs into land-use decisions. Proc. Natl. Acad. Sci. USA 2012, 109, 7565–7570.
9. Boody, G.; Vondracek, B.; Andow, D.A.; Krinke, M.; Westra, J.; Zimmerman, J.; Welle, P. Multifunctional agriculture in the United States. BioScience 2005, 55, 27–38.

10. Santelmann, M.V.; White, D.; Freemark, K.; Nassauer, J.I.; Eilers, J.M.; Vaché, K.B.; Danielson, B.J.; Corry, R.C.; Clark, M.E.; Polasky, S.; et al. Assessing alternative futures for agriculture in Iowa, U.S.A. Landsc. Ecol. 2004, 19, 357–374.

11. Waldhardt, R.; Bach, M.; Borresch, R.; Breuer, L.; Diekötter, T.; Frede, H.; Gäth, S.; Ginzler, O.; Gottschalk, T.; Julich, S.; et al. Evaluating today's landscape multifunctionality and providing an alternative future: A normative scenario approach. Ecol. Soc. 2010, 15, 30:1–30:20.

12. Price, J.; Silbernagel, J.; Miller, N.; Swaty, R.; White, M.; Nixon, K. Eliciting expert knowledge to inform landscape modeling of conservation scenarios. Ecol. Model. 2012, 229, 76–87.

13. Westervelt, J.; BenDor, T.; Sexton, J. A technique for rapidly forecasting regional urban growth. Environ. Plan. B: Plan. Des. 2011, 38, 61–81.

14. Patel, M.; Kok, K.; Rothman, D.S. Participatory scenario construction in land use analysis: An insight in the experiences created by stakeholder involvement in the northern Mediterranean. Land Use Policy 2007, 24, 546–561.

15. Alcamo, J. The SAS approach: Combining qualitative and quantitative knowledge in environmental scenarios. In Environmental Futures: The Practice of Environmental Scenario Analysis; Alcamo, J., Ed.; Elsevier: Amsterdam, The Netherlands, 2008; pp. 123–150.

16. Kok, K. The potential of fuzzy cognitive maps for semi-quantitative scenario development, with an example from Brazil. Glob. Env. Chang. 2009, 19, 122–133.

17. Walz, A.; Lardelli, C.; Behrendt, H.; Grêt-Regamey, A.; Lundström, C.; Kytzia, S.; Bebi, P. Participatory scenario analysis for integrated regional modelling. Landsc. Urban Plan. 2007, 81, 114–131.

18. Swetnam, R.D.; Fisher, B.; Mbilinyi, B.P.; Munishi, P.K.T.; Willcock, S.; Ricketts, T.; Mwakalila, S.; Balmford, A.; Burgess, N.D.; Marshall, A.R.; et al. Mapping socio-economic scenarios of land cover change: A GIS method to enable ecosystem service modelling. J. Environ. Manag. 2011, 92, 563–574.

19. Lamarque, P.; Artaux, A.; Barnaud, C.; Dobremez, L.; Nettier, B.; Lavorel, S. Taking into account farmers' decision making to map fine-scale land management adaptation to climate and socio-economic scenarios. Landsc. Urban Plan. 2013, 119, 147–157. Koschke, L.; Fürst, C.; Frank, S.; Makeschin, F. A multi-criteria approach for an integrated land-cover-based assessment of ecosystem services provision to support landscape planning. Ecol. Indic. 2012, 21, 45–56.

20. Hulse, D.W.; Branscomb, A.; Payne, S.G. Envisioning alternatives: Using citizen guidance to map future land and water use. Ecol. Appl. 2004, 4, 325–341.

21. Oregon Blue Book. City and County Populations. Available online: http://bluebook.state. or.us/local/populations/populations.htm (accessed on 21 August 2013).

22. Oregon Metro. 20 and 50 Year Regional Employment and Population Range Forecasts. Available online: http://library.oregonmetro.gov/files/20–50_range_forecast.pdf (accessed on 4 April 2013).

23. Portland State University College of Urban and Public Affairs Population Research Center. Population Estimates. Available online: http://www.pdx.edu/prc/population-estimates-0 (accessed on 19 August 2013).

24. Fry, J.A.; Coan, M.J.; Homer, C.G.; Meyer, D.K.; Wickham, J.D. Completion of the National Land Cover Database (NLCD) 1992–2001 Land Cover Retrofit Product. Available online:

http://pubs.usgs.gov/of/2008/1379/pdf/ofr2008-1379.pdf (accessed on 13 January 2014).

25. Fry, J.; Xian, G.; Jin, S.; Dewitz, J.; Homer, C.; Yang, L.; Barnes, C.; Herold, N.; Wickham, J. Completion of the 2006 National Land Cover Database for the conterminous United States. Photgramm Eng. Remote Sens. 2011, 77, 858–864.

26. Lambin, E.F.; Turner, B.L.; Geist, H.J.; Agbola, S.B.; Angelsen, A.; Bruse, J.W.; Coomes, O.T.; Dirzo, R.; Fischer, G.; Folke, C.; et al. The causes of land use and land cover change: Moving beyond myths. Glob. Environ. Chang. 2001, 11, 261–269.

27. Planning the Oregon Way: A Twenty-Year Evaluation; Abbot, C., Howe, D., Adler, S., Eds.; Oregon State University Press: Corvallis, OR, USA, 1994.

28. Oregon Department of Land Conservation and Development. Ballot Measures 37 (2004) and 49 (2007) Outcomes and Effects. Available online: http://www.oregon.gov/LCD/docs/publications/m49_2011-01-31.pdf (accessed on 1 April 2013).

29. Oregon Department of Land Conservation and Development. Metro Urban and Rural Reserves. Available online: http://www.oregon.gov/LCD/Pages/metro_urban_and_rural_reserves.aspx (accessed on 1 April 2013).

30. Boeder, M.; Chang, H. Multi-scale analysis of oxygen demand trends in an urbanizing Oregon watershed, USA. J. Environ. Manag. 2008, 87, 567–581.

31. Praskievicz, S.; Chang, H. Impacts of climate change and urban development on water resources in the Tualatin River basin, Oregon. Ann. Assoc. Am. Geogr. 2011, 101, 249–271.

32. Oregon Department of Environmental Quality. Tualatin Subbasin Total Maximum Daily Load and Water Quality Management Plan: Chapter 1—Overview and Background. Available online: http://www.deq.state.or.us/wq/tmdls/docs/willamettebasin/tualatin/revision/Ch0CoverExecSummary.pdf (accessed on 3 March 2013).

33. Pratt, B.; Chang, H. Effects of land cover, topography, and built structure on seasonal water quality at multiple spatial scales. J Haz. Mater. 2012, 209/210, 48–58.

34. Chang, H.; Thiers, P.; Netusil, N.; Yeakley, J.A.; Rollwagon-Bollens, G.; Bollens, S.M. Relationship between environmental governance and water quality in a growing metropolitan area of the Pacific Northwest, USA. Hydro. Earth Syst. Sci. 2014. in press.

35. Oregon Metro. RLIS Live, Geographic Information System Data. Available online: http://www.oregonmetro.gov/index.cfm/go/by.web/id=593 (accessed on 21 August 2013).

36. City of Dayton, Oregon. Planning Atlas and Comprehensive Plan. Available online: http://www.ci.dayton.or.us/vertical/sites/%7B0813AE62-E15F-4C65-858B-10DDF2ABA1FE%7D/uploads/Complete_Copy_3-31-11.pdf (accessed on 11 March 2014).

37. Oregon Spatial Data Library. Available online: http://spatialdata.oregonexplorer.info/geoportal/catalog/main/home.page (accessed on 21 August 2013).

38. USFWS Geospatial Services. USFWS National Cadastral Data. Available online: http://www.fws.gov/GIS/data/CadastralDB/index.htm (accessed on 21 August 2013).

39. Oregon Water Resources Department. Water Protections and Restrictions. Available online: http://www.oregon.gov/owrd/pages/pubs/aquabook_protections.aspx (accessed on 21 August 2013).

40. Oregon State Archives. Department of Land Conservation and Development. Division 33Agricultural Land. Available online: http://arcweb.sos.state.or.us/pages/rules/oars_600/oar_660/660_033.html (accessed on 27 September 2013).

41. U.S. Census Bureau. Profile of General Population and Housing Characteristics: 2010. Geography: Portland-Vancouver-Hillsboro, OR-WA Metro Area (part); Oregon, USA, 2010. Available online: http://factfinder2.census.gov/faces/tableservices/jsf/pages/pro-ductview.xhtml?pid=DEC_10_DP_DPDP1&prodType=table (accessed on 19 August 2013).

42. U.S. Bureau of Labor Statistics. Local Area Employment Statistics, U.S. Department of Labor. Available online: http://data.bls.gov/cgi-bin/dsrv (accessed on 19 August 2013).

43. Oregon Metro. The Nature of 2040: The Region's 50-Year Plan for Managing Growth. Available online: http://library.oregonmetro.gov/files/natureof2040.pdf (accessed on 28 January 2014).

44. Environmental Science Research Institute. ArcGIS 10.1.; Environmental Science Research Institute: Redlands, CA, USA, 2010.

45. Saaty, T.L. The Analytic Hierarchy Process: Planning, Priority Setting, Resource Allocation; McGraw-Hill: New York, NY, USA, 1980.

46. Castella, J.; Trung, T.N.; Boissau, S. Participatory simulation of land-use changes in the northern mountains of Vietnam: The combined use of an agent-based model, a role-playing game, and a geographic information system. Ecol. Soc. 2005, 10, 27:1–27:32.

47. Rounsevell, M.D.A.; Ewert, F.; Reginster, I.; Leemans, R.; Carter, T.R. Future scenarios of European agricultural land use II. Projecting changes in cropland and grassland. Agric. Ecosyst. Environ. 2005, 107, 117–135.

48. Parker, D.C.; Manson, S.M.; Janssen, M.A.; Hoffman, M.J.; Deadman, P. Multi-agent systems for the simulation of land-use and land-cover change: A review. Ann. Assoc. Am. Geogr. 2003, 93, 314–337.

49. Kok, K.; van Delden, H. Combining two approaches of integrated scenario development to combat desertification in the Guadalentín Watershed, Spain. Environ. Plan. B: Plan Des. 2009, 36, 49–66.

50. Southworth, J.; Marsik, M.; Qiu, Y.; Perz, S.; Cumming, G.; Stevens, F.; Rocha, K.; Duchelle, A.; Barnes, G. Roads as drivers of change: Trajectories across the tri-national frontier in MAP, the southwestern Amazon. Remote Sens. 2011, 3, 1047–1066.

51. Kelley, S.; (Washington County Land Use and Transportation Department, Hillsboro, OR, USA). Personal communication, 15 November 2012.

52. Radeloff, V.C.; Nelson, E.; Plantinga, A.J.; Lewis, D.J.; Helmers, D.; Lawler, J.J.; Withey, J.C.; Beaudrey, R.; Martinuzzi, S.; Butsic, V.; et al. Economic-based projections of future land use in the coterminous Unites States under alternative policy scenarios. Ecol. Appl. 2012, 22, 1036–1049.

53. Suarez-Rubio, M.; Lookingbill, T.R.; Wainger, L.A. Modeling exurban development near Washington, DC, USA: Comparison of a pattern based model and spatially-explicit econometric model. Landsc. Ecol. 2012, 27, 1045–1061.

54. Yee, D.; (Oregon Metro, Portland, OR, USA). Personal communication, 11 July 2013.

55. Walker, P.A.; Hurley, P.T. Planning Paradise: Politics and Visioning of Land Use in Oregon; The University of Arizona Press: Tucson, AZ, USA, 2011.

56. Peterson, G.D.; Cumming, G.S.; Carpenter, S.R. Scenario planning: A tool for conservation planning in an uncertain world. Conserv. Biol. 2003, 179, 358–366.

CHAPTER 9

The Use of Visual Decision Support Tools in an Interactive Stakeholder Analysis—Old Ports as New Magnets for Creative Urban Development

Karima Kourtit and Peter Nijkamp

9.1 INTRODUCTION

Ports are the oldest logistics centers in international trade and have always been economic powerhouses in transport systems. Already in the 18th century, the grandfather of modern economics, Adam Smith, was referring to seashores and riverbanks as poles of economic wealth, as their openness allowed them to establish trade relationships with the rest of the world. In the course of time, ports have developed as major logistic magnets generating trade and transport connections all over the world. In addition, consequently, many port areas laid the foundation for an increase in welfare, not only for the direct urban or industrial areas concerned, but also for the hinterlands connected with these areas, and for all other places served by these ports. Over the course of time, port areas have become hotspots of economic activity. The ports' history, culture and economy originate predominantly from their adjacent oceans, seas, lakes and rivers. They acted as (centripetal and centrifugal) transportation hubs that favored openness in trade in a global economy. They were also often the scene of socio-economic inequality, with a strong tension between white-collar managers ('barons') and blue-collar workers. Especially with the advent of the Industrial Revolution (mid-19th century), ports became symbols of a new industrial age, thanks to advanced steamships, large-scale shipyards, etc.

In the past few decades, many port areas all over the world went through a phase of decline, as they became outdated, or were replaced by modern facilities elsewhere. This has left many cities with large harborfront areas that were dilapidated and showed clear signs of environmental decay and even poverty. Such brownfield sites have increasingly become a source of policy concern, and have stimulated the emergence of various urban land-use initiatives in order to exploit the hitherto unused economic, social, logistic, cultural and environmental opportunities of such areas. As a result, in recent years many cities have developed new policy mechanisms for upgrading their port brownfield sites through harborfront and seafront development (e.g., the London Dockyards, the Kop van Zuid in Rotterdam, and waterfronts in Cape Town, New York, Yokohama, Singapore, Helsinki, etc.). The two key phrases in this drastic land use conversion are: sustainable development and creative sector stimulation. Port areas may thus become precious containers of past architectural and socio-cultural heritage and expression. But this heritage is not a passive phenomenon, but may be the basis for innovative developments in urban areas by offering new residential, business, and tourist facilities in the short and middle to long term.

Harborfront development and port revitalization are all part of urban gentrification processes. This has not only a physical dimension (e.g., land use, real estate, infrastructure), but also a socio-economic dimension, (e.g., labor force participation, inclusion of less privileged groups, cultural diversity). We refer in this context to relevant studies on these issues by Atkinson [1], Butler [2] or Watt [3]. Harborfront have many things in common, but they may differ in terms of their social and economic functions and activities, where important aspects related to urban renewal and revitalization—including policy scenarios for creative and sustainable urban development—are considered to be necessary for effective interventions. This means that a reformulation of port cities' policies may require 'out of the box thinking', while bringing together different perspectives, original interpretations, imagination, and appropriate tools for conflict management. This calls for imagining a future with a full understanding of the consequences before creating it. This may lead to a new process of (re)designing port cities ranging from small to large interventions regarding the functionality and architecture of the port areas concerned—interventions which preserve historical heritage (tangible and intangible) in combination with smart modern buildings and facilities. This task has to be undertaken against the background of a complex urban carousel of challenges in order to enhance the socio-economic and ecological resilience of the port area—in relation to the city system—and to activate many initiatives that would convert historico-cultural urban port landscapes into sustainable and creative hotspots, starting by reusing, recovering, and regenerating such areas. However, the high degree of tension between different stakeholders' needs and local government strategies or urban planning initiatives related to these port areas may frustrate sustainable development. This presents the challenge to review the port city system from the perspective of new paradigms, based, for example, on 'creative minds' principles [4]. For the definition of creative minds, we use the classification created by TNO [5] as a basis for the interviews. The definition of all these branches of the creative industries is based on the standard industrial classification [6] from Statistics Netherlands (CBS), which contains three types of creative firms, viz. arts, media and entertainment, and creative business services. Table 1 shows which economic activities are classified in these three groups.

Such novel ways of thinking are increasingly required and linked to new, efficient and effective urban planning, governance, and management processes in order to finally ensure broad stakeholder acceptance. This calls for new evaluation and assessment tools and techniques in order to confront decision makers with a consistent set of sustainable strategic choices and changes (by extending the range of possibilities and preferences) in relation to creative urban development, while considering a stakeholder participation-based approach ('bottom up' strategy) [7,8], using, for example, interactive methods based on what we call the 'urban Facebook' concept. In this context, such a concept could be an important analytical vehicle, through a multilayered bottom-up approach, to systematically map out various local stakeholders' needs, knowledge domains, interventions and perspectives (backcasting) into new visions and urban development strategies [9]. These various stakeholders may be characterized as early adopters or social innovators [10], and hence their behavior may be seen as trend-setting for a much larger population.

The present paper aims to further develop the multilayered stakeholder-based framework by introducing and elaborating what we call the 'urban Facebook for urban facelifts', an approach that is extensively supported by high-quality visual assessment tools for mapping novel redevelopment initiatives, in order to be able to identify and understand specific local needs and necessary spatial developments. It offers a basis for interventions that tackle present and future urban problems, foundations, challenges, and consequences (in combination with urban scenarios, which are essentially strategic future image experiments based on, for example, the imagining of future port cities' positions) designed to achieve the desired goals related to urban strategic visions, while, in addition, this approach may encourage the urban economy to stay (internationally) competitive.

TABLE 9.1   The classifications of the stakeholders 'creative minds' in the creative industries and the SBI codes: Arts, Media, and Creative Business Services (Source: [4]).

Main domains	Segments	Standard Industrial Classification (SBI)	
		SBI-1993	Description
Art	Music & Performing Arts, Museums, Theatres and Art galleries	92311	Performing of live stage art
		92312	Production of live stage art
		92313	Performing of casting art
		92321	Theatres, concert rooms, concert buildings
		92323	Services for performing art
		92521	Art galleries, exposition areas
		92522	Museums
Media	Film, TV, Radio, Photography, Publishing Broadcasting, Amusement and entertainment, Press	2211	Publishers of books
		2212	Publishers of periodicals
		2213	Publishers of magazines
		2214	Publishers of sound recording
		2215	Other publishers
		74811	Photography
		92111	Production of movies
		92112	Supporting services for movie production
		92201	Broadcasting organizations
		92202	Production of radio- and TV programs
		92203	Supporting activities for radio and TV
		9212	Distribution of movies
		9213	Cinemas
		92343	Other entertainment
		9240	Press-, news agencies; journalists
Creative Business Services	Advertising and Marketing, Information and Technology, Architecture, Design and Fashion	74201	Architecture and technical design
		74202	Technical design/advice, e.g., city building
		74401	Commercial design- and consultancy agencies
		74402	Other commercial services
		74875	Interior and fashion designers

Our empirical case study was carried out in and around the NDSM-area, a former dockyard in Amsterdam, the Netherlands. This study was undertaken in the context of transitional urban port systems for sustainable urban development, from a forward-looking long-term strategic policy perspective (a combination of backcasting and forecasting approaches), which meets the needs, and addresses the concerns, of its various users, where vision and strategy have to fit well with their environment. This bottom-up approach is, inter alia, based on information collected, during interviews, from different stakeholders with a wide range of interests in relation to the area, followed by the use of a strength-weakness opportunities-threats (SWOT) analysis methodology with visual support tools. All this is done in order to develop a collective and quantitative evaluation of the socio-economic performance of the NDSM, which focuses on its physical use, characteristics, and historical landscape attributes.

This paper is organized as follows. Section 2 is devoted to an overview of the NDSM Wharf as a new development core in Amsterdam. Then, Section 3 presents the methodology for assessing the NDSM-district, whereby past, current and future effects are assessed from a broad perspective. Section 4 then introduces the SWOT analysis, which leads to the design and presentation of a conceptual 'urban Facebook' for an NDSM Facelift. In Section 5, the urban Facebook is elaborated, including the database employed and the architecture of our exploratory data analysis, leading ultimately to the identification of the most suitable 'Urban Facelift'. This information is then further used in Section 6, which presents the results of our operational 'urban Facebook' by linking the present performance of the area to various future perspectives and urban future images for the revitalization of the NDSM. Finally, Section 7 makes some retrospective and prospective observations on our policy research, in particular on the NDSM district.

9.2   THE NDSM WHARF AS A NEW DEVELOPMENT EXPERIMENT

The NDSM, a former dockyard on the northern banks of the River IJ in Amsterdam, has become a culture-based creativity and social innovation district with a great diversity of trendy facilities and seemingly uncontrolled land use. This area is sometimes called the NDSM-Safari (see Figure 1).

The 'NDSM-Safari' serves as a laboratory of informal and formal living and working spaces with new infrastructure for the collaboration of creative minds—a bubbly mix of activities directly involved in the development and production of cultural, creative and innovative products and services—for new urban development and advanced urban competitiveness. It is hoped that creative minds will develop innovative ideas and suggest new pathways to sustainable development, and act as central breeding places for a broad range of various stakeholders in search of original concepts in a globalizing competitive world [4]. From this perspective, creative minds have an exceptional innovation potential in terms of both ideas and practices. Therefore, they may act as effective growth engines in modern cities.

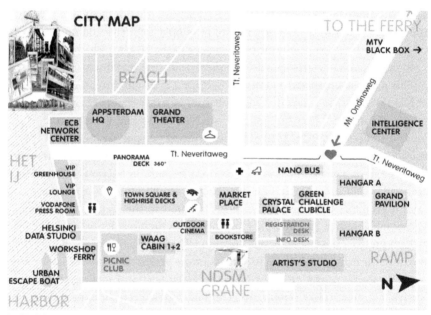

FIGURE 9.1   Map of the NDSM-Safari in Amsterdam.

The existing 'bohemian landscape' of the NDSM shipyard, with its historical background, exploits its rough and untouched diversity and flexibility characteristics. It has gone through a number of different phases generated by the creative minds of the district and renews itself often, so its creative bubble of mixed functionality and working class population [11] is well-recognized and a great inspiration for the next stage of the development of the urban core. It can provide many illustrations of an informal repositioning and redevelopment of a district with a great potential, but the local authority has no clear long-term strategic view or commitment regarding the potential of the 'creative minds'.

The future form of the NDSM-Safari is already taking shape. However, as just mentioned, it lacks both long-term strategies (e.g., a solid and integrated breeding place policy) to meet the important needs and preferences of the various stakeholders and guarantees by the local authority to create a 'sustainable home' for various professionals, businesses, and artists. And it should not remain only as a temporary 'project', but become a new part of a future productive urban landscape instead of an isolated breeding place. This means that viable strategic options have to be interpreted and discussed in an integrated multilayered framework in order to provide a sound basis for the possible preparation of conditions for the further redevelopment of the NDSM location as a district for the production of urban culture. There, place-based characteristics and opportunities, and historical landscape attributes may draw (more) creative minds and innovative business models to certain sites, where they can share and combine their (international) knowledge and expertise with challenging socio-economic opportunities. This requires an understanding of more than just the commercial side of this district or the decrease of the 20 sub-clusters located there, in order to realize their common interests in the NDSM vision and come to a general strategic core policy.

It is noteworthy that the presence and experience (individual visions, preferences and values) of creative minds can create critical conditions for the level of attractiveness of this historical and cultural district as a favorable concentration of geographical space (clusters). In and around the NDSM district, the various stakeholders can experience the inspiring urban atmosphere and 'cool image' (e.g., visual features, reputation), which are crucial for creative and innovative working processes. The value, for instance, that firms put on the NDSM-district regarding their location-decisions based on their preferences intensities and criteria for visual assets may positively influence these firms' (business) performance and strategic choices. These, in turn, may bring about positive socio-economic achievements, which may enhance the attractiveness of port cities and regions and, ultimately, achieve a high degree of sustainability and competitive advantage [12]. In order to generate positive externalities, regions and cities have to listen to the various stakeholders and provide unique geographical and location conditions and facilities—beyond other competitive assets—in order to attract talent and firms to relatively deprived regions. This issue has been repeatedly addressed in the past by the local authority.

From this perspective, the present situation regarding the sustainable development of NDSM calls for a careful evaluation of this hotspot, which assumes that this 'port system within a system' of the city Amsterdam [13] will create the possibilities and new opportunities for the entrance of new cultural and innovative activities such as 'hip' cultural areas, which can give a sense of freedom to (new) creative minds. Hence, the main aim of this empirical research is to develop and support a new and promising future orientation for the district's sustainable development, based on a multilayered stakeholder-oriented approach (which, in redesigning the port area, attempts to resolve conflicts between the interests or values of a multiplicity of stakeholders, while favoring economic prosperity in combination with meeting social needs).

9.3   RESEARCH METHODOLOGY

Our central methodological research task is to develop a multilayered stakeholder-based analysis framework that is fit-for-purpose for the NDSM-district, and is able to analyze its future potential, using interactive methods, where the form of the urban facelift embraces different levels of urban revitalization for the district's sustainable development.

This conceptual evaluation framework can be characterized as a pro-active process of choosing viable strategic choice(s) and a facelift for areas based on various stakeholders' preferences and values, in the form of a SWOT analysis. In this analysis, the indicators that can influence the constellation of socio-economic characteristics of port and city systems were converted into a long list of important criteria and presented to the interviewees. This process can be systematically divided into six steps, with regard to the socio-economic context (based on the 'Strategic Choice Analysis' (SCA) approach [14], starting from the NDSM development assessment, and ending with conclusions and the identification of policy recommendations for a new urban facelift (see Figure 2). This approach can also be seen as a toolkit for the sustainable development of other port cities, cities and regions, which are aiming to enhance their dynamic profile.

This process starts with the assessment of the physical use, characteristics, and historical landscape attributes and developments of the NDSM district, and ultimately identifies general strategic ideas for its future sustainable development (by backcasting and forecasting). Furthermore, this process is fully supported by a collection of interesting pictures of buildings, restaurants, hotels, abandoned areas, the general atmosphere, and spaces (urban faces) that can be considered as important factors for the district's future sustainable development.

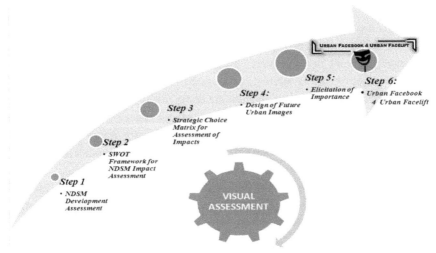

FIGURE 9.2   Stepwise presentation of the evaluation to identify the most effective strategic choices, images, and opportunities for the sustainable development of the NDSM district.

Based on these most representative pictures of the NDSM-district, which illustrate both positive and negative aspects, four representative urban faces emerged, and were presented to a group of important stakeholders (from artists to entrepreneurs), operating in and around the NDSM-district, during a semi-structured interview. This was done in order to find the optimal level of revitalization to maximize their strategic options and opportunities for future sustainable development. Figure 3 presents the four strategic urban faces of the NDSM district as they exist at present.

FIGURE 9.3   The four most representative pictures of the NDSM-district.

These four strategic urban faces are based on two dimensions: first, the combination of historical heritage (tangible and intangible) and modernity, regarding the functionality and architecture of the district; and, second, either piecewise (project) or integrated (program) redevelopment of the district. From this point of view, a 'strategic future urban faces diagram' can be created in which specific levels of revitalization are distinguished (see Figure 4). These four strategic urban faces cover both local and global scales.

To evaluate the performance of these four strategic urban faces, each stakeholder was asked to rank each urban face with a score, varying from '1 = low' to '5 = very high', according to a long list of criteria (pairwise comparison of indicators) (see Appendix A, Table A1) extracted from the SWOT analysis. Interviewees were also asked if there were other important indicators they have had experienced. The long list of criteria gathered from extensive interviews with various stakeholders and literature sources on, respectively, the negative and positive benefits of the NDSM area (in a SWOT analysis) was then systematized and summarized by means of a Principal Component Analysis (PCA). This enabled us to extract the four most important strategic components: namely, economic vitality; accessibility; cultural diversity; and ecological sustainability (for details of the PCA see Appendix B, Tables B1 and B2 and Figure B1).

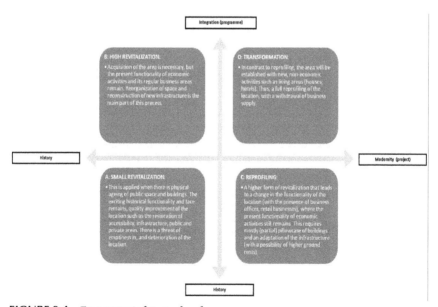

FIGURE 9.4   Four strategic future urban faces.

To link the long-term strategies to short- and medium-term operations, these strategic domains have to be translated into a new potential vision of the NDSM district that can lead to a clear strategic direction and effective related actions. Thus the focus on these strategic domains needs a clear interdisciplinary orientation that is centered on the future sustainability of the port city system, which may bring breakthrough innovations on that could reinforce the creative bubble of mixed activities and multifunctionality in and around the NDSM district.

Table 2 shows the long list of indicators that we used to evaluate each of the four components with regard to the values and preferences of the four strategic urban faces. The NDSM-district calls for strategic public governance systems that reinforce its potential. To assess the 'competitive advantage' a la Porter of

TABLE 9.2 Evaluation indicators of the four components with regard to the values and preferences of strategic urban faces.

Main domains	Criteria
Accessibility & Learning School	*Independent ultimate events venue* *Enjoyment* *Dynamic 'oasis'* *Accessibility* *Independence & creative atmosphere* *Learning* *Function* *Transportation*
Innovation & Economic Vitality	*Urban socio-economic climate* *Traditional workspaces and activities* *Cultural profile* *Creative image* *Strong cultural and creative profile* *Business climate* *Long-term strategies* *Quality of life and sustainability*
Cultural Diversity & Entrance	*Quality of urban life* *Demography* *Cultural amenities* *Low rent*
Quality & Ecological Sustainability	*Urban design and architecture* *Urban land use* *Criminality*

SOURCE: authors' elaboration.

such new urban governance systems, it is necessary to design a relevant indicator system that shapes the multilevel creative resources of this urban district. This is also a necessary step for a benchmark performance analysis of the success and failure conditions of urban policy. Clearly, such indicators should be transparent, manageable, testable, comparable, representative, and policy-relevant [15].

Next, to extract from these main domains systematic and coherent viable long-term strategies for sustainable urban development, we adopted four related thematic alternatives, which we called 'urban future images' of stylized appearances of urban agglomerations in the year 2050, introduced for the first time in Nijkamp and Kourtit [16]. These urban images may be used as strategic vehicles to identify important challenges and foundations for the innovative development of the NDSM district towards a new 'urban facelift' (by forecasting and backcasting approaches). These strategic alternatives are briefly described in Table 3. Thus, to evaluate the performance of these four strategic urban faces (A–D), each stakeholder was asked to rank each urban face with a rating, varying from '1 = low' to '5 = very high' in the context of the four alternative 'urban images 2050'—in a structured impact matrix, where the alternatives refer to future developments of the NDSM district.

The images in this Facebook are based on smart-physical and immaterial-infrastructure. All of their elements are centered on a spatially integrated force field for the NDSM-area that entrances the competitive capacities of different stakeholders in that area. In a recent study [15], various analytical contributions can be found, such as the FIRES-Quare model, the XXQ Pentagon model, the leader and organizing capacity approach, or the smart infrastructure model.

This 'urban Facebook' framework has found support from previous scientifically orientated works that have shifted the focus of evaluation models in policy design and urban planning towards the sustainable development of the diversity of the important values and preferences relating to urban areas. This includes the need for the involvement of different stakeholders whose preferences and values are associated with these areas, in the process of design, and ultimately implementation. Examples of such works are: 'strategic choice analysis' [14]; the evaluation of historical districts in cities [17]; and the assessment of district visual quality in the location decisions of creative entrepreneurs [18].

This approach improves and increases the ability to recognize the importance of understanding the characteristics of an area and the preferences for socio-economic and environmental values, including the involvement of all stakeholders' interests in a way that brings and keeps them together, and thus offers a broader perspective regarding the district's sustainable development. It

TABLE 9.3 Strategic Choice Impact Matrix.

	FUTURE URBAN IMAGES			
	The Entrepreneurial City 2050	The Connected City 2050	The Pioneer City 2050	The Liveable City 2050
Urban Faces Urban face A Urban face B Urban face C Urban face D	This image assurances that in the climate of current and future global and local competition, Europe can only survive if it is able to maximize its innovative and commercial potential in order to gain access so emerging markets outside Europe; cities are then spearheads of Europe's globalization policy. This image covers a range of entrepreneurs, from SME's and migrant entrepreneurs to globally-operating firms. The key drivers behind this image, and supporting the change, are innovation and economic vitality.	The image of a connected city refers to the fact that in an interlinked (from local so global) world, cities can no longer be economic islands in themselves ("no fortresses"), but have to seek their development opportunities in the development of advanced transportation infrastructures, smart logistics systems and accessible digital communication systems, through which cities become nodes or hubs in polycentric networks). The key drivers that influence the form of this image are Smart logistics and sustainable mobility.	This image refers to the innovative "melting pot" character of urban areas in the future, which through open communication channels will show an unprecedented cultural diversity and fragmentation of lifestyles in European cities; this will present not only big challenges but also great opportunities for smart and creative initiatives in future cities, through which Europe can become a global pioneer. In this image important key elements to reduce the welfare gap are social participation and social capital.	This ecological-based image addresses the view that cities are not only energy consumers (and hence environmental polluters), but may through smart environmental and energy initiatives (e.g. recycling and waste recuperation) act as engines for ecologically-benign strategies, so that cities may become climate-neutral in a future space-economy; cities in Europe will them be attractive places to live and work, with a global international outreach. In this image, the critical components ecological sustainability and high quality of between resources and efforts used to achieve his.

is noteworthy that the local authority has to realize that it needs the support of important stakeholders (private companies and, for instance, representatives of civic organizations) to make the revitalization and the implementation of urban facelifts successful. It is of the utmost importance to distinguish the various stakeholders, and involve them in the planning process of sustainable development under uncertainty.

9.4   NDSM STATE OF THE ART: SWOT ANALYSIS

To position the impacts on the NDSM-district in a broader strategic context of socio-economic benefits, this section give a systematic overview of the various effects, mainly in the form of a Strength-Weakness Opportunities-Threats (SWOT) analysis, in which past, current, and future effects are assessed from a broad perspective, extensively supported by the strategic urban faces (A–D) [14]. This review results in the construction of a list of the impacts on the NDSM-district that is used here as a case study. In this connection, the stakeholders were asked to identify and to prioritize the most important strength (S) and weakness (W) factors for the NDSM-district from a long-term strategic perspective for the distinct domains of innovation development and cultural diversity importance (derived from the strategic view concerning the future of the NDSM-area).

The results in Table 4 and Table 5 show the key factors, as identified by the various stakeholders in the process, including both the S and W elements. These data indicate the relevant factors of both domains of the NDSM district, along with their impact on the elements Opportunities (O) and Threats (T). This information represents the vital and creative contribution of creative minds to the urban economy, and can aid the development of appropriate strategic policies for countries [14]. Table 4 and Table 5 present the impacts from the SWOT analysis undertaken in this section, and will be used later in this paper as an input in the framework developed for identifying the level of revitalization necessary to achieve a new urban facelift and related strategies with regard to the NDSM-district.

Table 4 shows that the majority of the stakeholders valued innovation force and creative industry (S1 and S2) as the most important strengths, which have a strong impact on new products, new markets, urban vitality and also the creative business climate (O1, O2, O3 and O4). Where below average growth and transfer abroad (W1 and W2) were identified as the most important weaknesses, these have a strong impact on poor institutionalization, the rise of the informal economy (T1 en T2), and urban vitality (O3).

TABLE 9.4   SWOT analysis—innovation development.

Innovation Development	
Strengths (S)	**Weaknesses (W)**
1. *Innovation force****	1. *Below-average growth****
2. *Creative industry****	2. *Transfer abroad****
3. Innovative cluster	3. Recognition of creative minds
4. Entrepreneurship	4. Poor professionalization
5. Strength of competition	5. No long-term strategies or clear vision
6. Economic growth	6. Youth participation
7. Supply of affordable work and living spaces	7. Traditional sectors
Opportunities (O)	**Threats (T)**
1. *New innovative products and services*	
2. *New markets*	
3. *Urban vitality*	
4. Launch new initiatives	
5. Employment opportunities	1. *Poor institutionalization****
6. Creative business climate	2. *Rise of the informal economy****
7. Internationalization of the city	3. Temporary area projects
8. Knowledge spillovers	
9. International contacts	
10. Enhancement of small-& medium-sized businesses	
11. Sustainable competition	

Source: author's elaboration.

Table 5 shows that the majority of the stakeholders were convinced that creativity and a strong economic profile of businesses (S1 and S2) are the most important strengths, and both have a strong impact on innovativeness (O1) and neighborhood criminality (T1). However, quality of life and the dual society (W1 and W2) are experienced as the most important weaknesses, both having a strong impact on social solidarity, social cohesion (O2 en O3), and neighborhood criminality (T1).

TABLE 9.5   SWOT analysis—cultural diversity.

Cultural Diversity	
Strengths (S)	**Weaknesses (W)**
1. *Creativity****	
2. *Economic profile****	
3. Network organizations well-know and connected with the Dutch and international creative industries	
4. Urban benefits 'cultural diversity'	1. *Quality of Life****
5. Socio-cultural enrichment	2. *Dual society****
6. Strong social networks	3. Insufficient use of cultural diversity
7. Diverse facilities	4. Lost of trust in the government
8. Accessibility	5. Abandoned areas
9. Cultural and free image	6. No structured interaction between
10. High quality lifestyles	creative minds and the local authority
11. Cultural identities	
12. Connect informal and formal networks	
13. Largest cultural hub in Amsterdam	
Opportunities (O)	**Threats (T)**
1. *Innovativeness****	
2. *Social cohesion****	
3. *Social solidarity****	
4. A wide arrangement of resources and efforts	1. Neighborhood criminality
5. Inspiration and cultural expression	
6. Cooperation between different disciplines	
7. Reinforcement of internationalization	

SOURCE: author's elaboration.

In recent years, the creative industries [19] have received increasing attention from policymakers in the Netherlands, particularly in Amsterdam, where long-term policies for these industries are included in strategic city policies and the planning of several different fields. Amsterdam is the base for a rich diversity of cultural and economic activities, including international-related knowledge-intensive activities. It has developed many policy strategies that aim to attract the firms of the creative industries, especially SMEs, in order to encourage the further development of this promising sector for socio-economic growth. Abandoned industrial locations like the former NDSM shipyard are being gradually transformed stage by stage into attractive locations as 'creative and innovative hubs' with a 'cool' image for a growing number of talented and skilled firms and people in both the creative and other industries.

Nowadays, the NDSM district is the largest cultural hub in Amsterdam, and offers facilities for several artistic disciplines. Thus, the area is not only a geographic hub for the bohemians, but is also becoming a strong 'innovative cluster' for various firms in the creative industries. Over the years, it has become the place for the creation of employment opportunities and the supply of affordable work and living spaces, and it has presented a cultural and unconstrained image for potential users of the district. All this plays an important role in the development and maintenance of high-quality lifestyles and cultural identities within the city. It brings together several informal and formal networks that create added value (incl. inspiration, cultural expression opportunities, cooperation between different disciplines). Furthermore, it encourages innovation forces in the creative industries and the launch of new initiatives (e.g., PICNIC, CCAA, Amsterdam Creativity Exchange, Amsterdam Innovation Motor, HTNK), new products, and the development of new market segments (most artists cannot live from the art market alone), which all contribute to the attractiveness and vitality, and an increase in the national and international appeal of the port and city system. Already in 1989, Harvey stated that "it particularly does so when an urban terrain is opened for display, fashion and the 'presentation of self' in a surrounding of spectacle and play. If everyone, from punks and rap artists to the 'yuppies' and the haute bourgeoisie can participate in the production of an urban image through their production of social space, then all can at least feel some sense of belonging to that place" ([20], p.13).

Unfortunately, the cultural diversity of this 'urban terrain' in Amsterdam is insufficiently used as a cultural, economic and international asset for the broadening and growth of the creative industries, which could put the area on the international map. In the opinion of the various actors, the importance and presence of these available innovative businesses, and their trendy products and services for the various market (mostly highly segmented) that provide future income sources, have not been sufficiently recognized by the public and the government.

The creative professionals do often not interact with the local authority on a structural basis. However, the new initiatives (network organizations) are well known and have connections with the Dutch and international creative industries both to stimulate their maturity and professionalization and to improve their economic performance. They open their informal and formal networks for the creative industries in order to develop a coaching and managing trajectory (a one-stop-shop for the creative industries) and increase the wide provision of resources and efforts that play a very important role in halting the neglect of areas. Furthermore, by enhancing their cultural position, realizing a varied and high-quality image (the general atmosphere and hospitality), bundling and strengthening their innovative powers, increasing national and international knowledge sharing and contemporary collaborations of different partners and firms (social cohesion and social solidarity), they help to turn a raw talent into a potential and professional entrepreneur and attract international sponsors and large projects (e.g., the Red Light Fashion Amsterdam Project, Redlight Design Amsterdam) for establishing cross-cultural collaborations, creative and innovative networks and (formally) strengthening Amsterdam's international cultural reputation and the city's economic growth (and the decline of the informal economy in this district).

It would help to invest more in the development of creative and cultural competences and talents (entrepreneurial skills) and to connect the chain of cultural and economic activities, as the NDSM-district could stimulate high quality cultural production and international trade-connections. Therefore, it is important to: create opportunities to stimulate young creative and innovative entrepreneurs; encourage their creativity, professional development (economic independence) and entrepreneurship; market their activities more efficiently; and innovate in order to adjust to changing markets, both locally and internationally. Therefore, opportunities to share knowledge, exchange information and gain access to international market segments are very important. In this respect, the interdisciplinary engagement of innovative professionals from different fields, such as architecture, music, and art, would provide an opportunity to develop innovative products.

In conclusion, the NDSM-district is a good example of a source of creativity and innovation, but is still lacking a clear and robust vision to develop a 'sustainable home'. This increases the risk of losing important of growth opportunities to transform or improve the district and city's image into a cultural and creative 'safari' that clearly reflects its unique identity and high values, including the involvement of creative minds and their preferences in order to develop a shared vision that illustrates a strong synergy of collective expertise and development plans. This new approach leads to new opportunities for different levels of revitalization, in the transformation from a historical industrial area into a more healthy creative urban bubble of mixed functions. Here, the innovative hub with its cultural activities serves several more goals than just providing a cheap place to work and live. Moreover, it clearly stimulates a social infrastructure or network and contributes to the cultural enrichment of a city.

The next section takes a closer look at the possibility to have an 'urban facebook for urban facelifts' for the NDSM shipyard, based on the preferences of creative minds and the value they put on the NDSM district.

9.5 AN URBAN FACEBOOK FOR FACELIFTS

The SWOT analysis resulted in an overview that respects the history of the area and integrates many of the historic elements, in order to identify its potential 'look' and find the optimum solution for the area by taking into consideration the stakeholders' opinions and points of view. Based on the SWOT analysis and the strategic and visual assessment derived from the interviews, our Strategic Choice Analysis (SCA)—as a vehicle for assessing and developing strategic policies for the development of the NDSM district—aims to identify [14]:

- the most important Strength factors to be used to participate in, or take advantage of, Opportunities (SO strategies) and to counter or avoid Threats (ST strategies) with regard to the various levels of revitalization;
- the most important Weakness factor to be eliminated (SO) or improved in order to be able to participate in Opportunities (WO strategies) and to counter or avoid the impact of Threats (WT strategies) with regard to the various levels of revitalization;
- the most important strategic proposals to be used to take advantage of the Strengths and Opportunities detected or to avoid the Weaknesses and Threats identified with regard to the various levels of revitalization.

The results are organized according to the importance of the NDSM landscape, future strategies, the needs, preferences, and values of the users, and the safari-profile defined as policy guidelines for the sustainable development of the area. To determine the degree of importance of the various levels of revitalization, the rank order of the urban faces ranges from 0 points for each irrelevant impact to 5 points for the most important impacts. After multiplying each score with its given importance classes, we are then able to synthesize all scores to determine the strongest factors for the two relevant, socio-economic domains derived from the strategic view for the NDSM district. Once all the important factors have been reviewed in order to assess perceived importance categories, strategic choices are then made by selecting the particular urban face that will most greatly influence policy strategies, viz. a combination of S and W elements for the two relevant socio-economic areas of the NDSM area, along with their impact on O and T. All this information can aid in the development of appropriate strategic policies for the port and city, including the NDSM district [14].

In Section 4, several challenges and opportunities were identified for the transformation and improvement of the NDSM-district, with a view to providing a balanced future for an XXQ urban system (based on the 'XXQ' principle that refers to the highest possible urban quality [21]). According to this, the unique NDSM area definitely needs a clear and transparent long-term strategy that takes into consideration both the entire potential of the place and the individual preferences, in order to ensure a high degree of transparency and future security. The spider diagram in Figure 5 represents the importance and values with regard to the levels of revitalization recommended by various stakeholders, according to the long list of criteria considered in the urban facebook evaluation system.

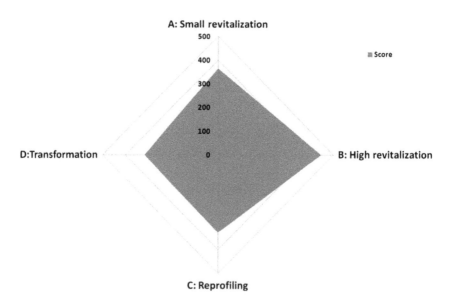

FIGURE 9.5   Visual representation of the various levels of revitalization recommended by the stakeholders.

The findings shown in Figure 5 indicate that, on the basis of the long list of criteria, Urban Face B that envisages a high degree of revitalization for the NDSM district is considered by the majority to be the most important and preferable strategic choice for it to become a place for a variety of events and cultural activities which can bring visitors and business into the area. The second preferred strategic option for the district is Urban Face A with a small degree of revitalization, where the presence of different users and socio-economic segments and activities in and around the district, including professional needs and interests will arise in the (re)development process. The results show that Urban Faces C and D compared with the other two Urban Faces play a less dominant role in the preferences.

Figure 6 shows that Urban Face A includes a combination of key effective forces, such as creative atmosphere (score: 19), learning (score: 19), accessibility (score: 18), independent events venue (score: 17), dynamic 'oasis' (score: 17), traditional workspaces and activities (score: 18), all of which are driving evolutionary and successful change, in an overall transition from a weak to a strong multi-functional and energetic creative hub.

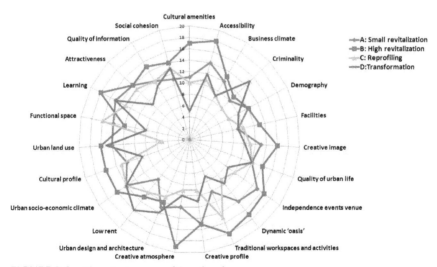

FIGURE 9.6   The criteria scores of the urban faces.

The NDSM area is a place of creativity and innovation (the main strengths of the area), and is currently being experienced as a 'warm nest for creative minds', and a place that provides the creative inspiration that the stakeholders need in order to develop their projects, and where they can develop their entrepreneurial skills without being disturbed. Furthermore, it is a perfect place for all kinds of events. The community consists of young creative pioneers—the average age of the entrepreneurs in the NDSM is very low, so the area can become a place for the young generation to start their business and enrich their professional experience (in sustainability learning centers supported by innovation forces). Thus, it is a 'welcome place' for start-ups where the creative and innovative community is consolidated and always willing to help. This presents a potential 'look' that maintains the district's important historical originality and architecture, but with a strong recommendation for the reorganization of space and the reconstruction of new infrastructure, while simultaneously preserving the cultural values.

However, the need for a long-term strategy for the area is still felt strongly among the creative minds, although they did not always agree with the previous transformations and direction of (re)development. For example, event planners would like to attract more tourists to this area, while the artists do not agree with this idea. However, they do share certain opinions based on their field of activities, and do believe that is necessary to strike a balance between independence (to give a feeling of creative freedom to various people about all the possibilities of the area) and efficient regulation and policies, and a balance between the need to revitalize the economic side of the area (tourism) and the possibility to be able to explore their creativity (core business) without constraints.

Finally, they still do not see a clear relation between the selected pictures and the criteria to score the performance of the Urban Faces in and around the NDSM-district. This brings our research study to the next level in the urban Facebook framework and a prompt breakthrough, based on the list of criteria, in developing a realistic future potential look ('Urban Face') and *a place 4 all*. The lack of clear long-term strategies and steering mechanism (governance strategies) may distort people's perception and points of view. This suggests the need to create a complex 'welcome image' regarding the history and the current use of the area in order to adopt an integrated approach for the realistic metamorphosis of the area—including the involvement of all stakeholders' interests, which brings and keeps them together and offers a broader perspective regarding the district's sustainable development.

In conclusion, the presence of the creative bubble of mixed functionality of cultural and economic activities, historical buildings and historic values, and memories is perceived as very important and represents a welcoming image of the area. However, the need to further revitalize the space and provide new infrastructure and networks is still an important part of this process. In this respect, the integrated approach of the local authority to the notion of 'creative community' is noteworthy. This embraces different sets of common preferences and options that together address issues in the evaluation process, and come up with the development of innovative solutions in order to prevent the area from declining in its attractiveness for the various stakeholders, who would subsequently experience a degradation in their quality of urban life (the 'XXQ' principle [19]); and, from a broader perspective, to ameliorate the socio-economic issues.

9.6   URBAN FUTURE IMAGES NDSM-DISTRICT

To evaluate the strategic position of the Urban Faces for the NDSM-district towards a place 4 all for future sustainable development, it is necessary to look at the present situation in terms of the performance across the different viable future image areas: the Entrepreneurial City, the Connected City, the Pioneer City, and the Livable City. These four future images highlight the strategic dimensions of urban futures in Europe. They lend themselves to systemic approaches for the future positioning of the NDSM-district, and reflect the need for strategic thinking on the governance of urban agglomerations for this area from different future perspectives. The ranking levels that each Urban Face achieves in a particular future urban image are: low (1); medium (2); high (3); and very high (4) (pairwise comparison) in order to identify which urban face fits best in different future circumstances in 2050. Figure 7 presents photos A–D in an order that best fits the importance of, and preferences for the four Urban Faces, in order to develop a shared vision and strategies that could lead to new opportunities for different preferences.

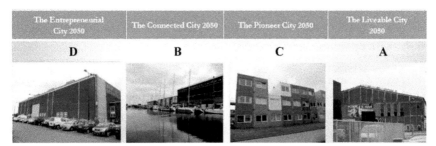

FIGURE 9.7   Urban faces positioned in best-fit urban perspectives.

9.6.1   Livable City 2050

Urban face A is strongly related to the Livable City 2050. This image shows a more or less abandoned area and buildings with no clear environmental function or relation to general city policy strategies for the coming years. However, it has strong historical value and special historical memories. Therefore, various stakeholders strongly prefer to keep its original structure and make it a liveable place for eco-tourists and visitors who visit the area for ecological reasons. This approach definitely needs high revitalization. This image perspective envisions the area as a place to live and for leisure activities, in other words, as a residential area.

9.6.2   Connected City 2050

This image focuses more on the connection between the NDSM-district and other parts of Amsterdam in terms of transport and infrastructure. This image refers more to the Connected City 2050, where just a small degree of revitalization is required. A few modifications are needed regarding this image to value its potential as a factor for the accessibility of the area.

9.6.3   Pioneer City 2050

This image needs a total transformation in order to keep the vibe ('spiky environment') in the area on a high level instead of only being a cheap 'sleeping place' for students—which leads us strongly to the Pioneer City 2050. However, students are a source of creativity and innovation, and need to be involved in this dynamic knowledge arena. This image is more connected to an entrepreneurial perspective that is open to new technologies and has the potential for innovation.

9.6.4   Entrepreneurial City 2050

In this case this area would be focused on creative business and start-ups. The building in Photo D and part of the area have an important meaning for all the people living and working in the NDSM district due to their history and activities during the weekend (a market has been opened inside). This area is not fully used. An option would be to give entrepreneurs more freedom and space to develop all kinds of activities in this area and inside the building so that the place may become more productive. According to various stakeholders, this Urban Face will need new functions, and therefore it has to be included in the reprofiling category.

In conclusion, the lesson from the previous analysis is that the local authority should involve various stakeholders and gain their trust in (re)developing a positioning strategy towards a high level of transparency, and a high level of productive environment that can positively affect issues such as social segregation, housing policy, infrastructure and logistics, environmental sustainability, urban land use, smart energy use, negative urban externalities, and the NDSM district's (international) competitive position. All this requires novel insights and policy strategies in order to make the future '*a place 4 all*'.

This would give the area the opportunity to have a new kind of urban facelift in the form of the *Welcome City 2050*. The new image addresses the view that cities in an open world have always been a place of cultural exchange. The urban multi-historical component plays a key role in determining the identity, diversity, and cultural richness of the port and city system. In this image, both the port and city become an attraction pole for a rising number of international creative minds and tourists. The area is culturally and ethnically located in a strong diverse metropolitan area. Historically, different groups and activities in and around the district have populated this area, in which these groups have actually been seeking spatial segregation to strengthen the cultural and creative identity of this district, as is often the case with immigrant groups.

There is strong collaboration between the creative entrepreneurs working in the NDSM-district, mostly based on technical issues and part of their product value chain. The very act of sharing a workplace and/or working with people full of passion regarding their work and being helpful when needed can play an important role in their positioning strategy. Furthermore, the industrial and rough nature of the area, the free spirit, and being close to people who have the same interests are also factors which are considered to be as important in the location-decision of these entrepreneurs and affects their productivity. Therefore, the potential 'look' to become a 'warm nest for creative minds' is a realistic mission to achieve in order to build an extraordinary community that strongly contributes to the city's sustainable development, and includes them in the long term strategies (see Figure 8).

This image of intense revitalization and transformation prompts the need for a new intervention in innovative developments with regard to cultural modernity, with a clear focus on spaces of social interaction and cultural integration, and the enlargement of the variety of facilities, through the high-quality urban design and equipment of those spaces in the process of the creation of *a place 4 all*. This will make Amsterdam once more a place where energy is generated, which is a continuous process. This new identity is central to the overall transformation of Amsterdam North, with the adaptation or transformation of an older waterfront into a contemporary creative bubble of mixed functions.

This image refers to the innovative 'melting pot' character of urban areas in the future. There will be an unprecedented cultural diversity and fragmentation of lifestyles in and around this district; this will present not only big challenges but also great opportunities for smart and creative initiatives in the future city, whereby it can become a global pioneer. This environment provides various opportunities and solutions to create a connection between artists, economic activities, citizens and government. In other words, a further democratic development of a city-in-a city, what we call the City 2.0, which provides the perfect opportunity to create a living lab in a cultural city of high quality.

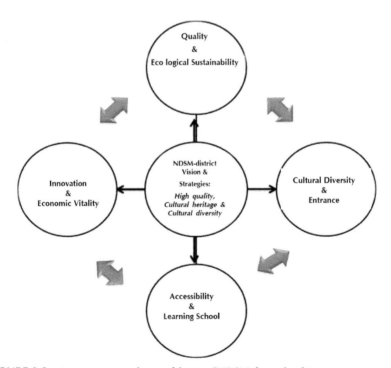

FIGURE 9.8   An interconnected view of the new "NDSM-district' architecture.

9.7.   DISCUSSION AND CONCLUSIONS

Clearly, cities are not only engines of economic progress, but are also places where cultural heritage is prominent. This also holds for port cities, which house a wealth of remains from the past: warehouses, silos, wharfs, lighthouses, industrial architecture, and so forth. It seems therefore, plausible to seek the anchor points of the urban rehabilitation of port areas in their undervalued land use related to logistic port activities from the past. It is noteworthy that the NDSM dockyard is being transformed further to have a more independent atmosphere with less support from the local authority, more successful and autonomous businesses, and ultimately an events venue (such as Robodock). It is becoming the largest cultural and creative project in terms of the city of Amsterdam's core breeding place policy.

Nowadays, port areas—even in a state of decay—often constitute the entry point and core area for the sustainable development of the entire urban system. Port areas offer an unprecedented heritage of a political, architectural, logistic, economic, social and artistic nature, with a great future potential. To understand and exploit this potential, it will be necessary to design an analytical framework which links the manifold opportunities provided by traditional port areas to future sustainable and creative urban development. This challenging objective needs a combination of forecasting and backcasting tools. From that perspective, there is a need to develop fit-for-purpose, dedicated policy tools and gentrification initiatives, on the basis of general planning principles for harborfront and seafront development. An ambitious implementation of policy goals associated with port development—such as job creation, foreign direct investment, creative sector development, environmentally-benign mobility, or sustainable land use—would thus become a major task for a modern port city. It will indeed be a great challenge to redesign and re-image port areas as multifunctional epicenters of creative urban initiatives and developments. These can essentially be seen as living laboratories and innovative urban areas for the development of sustainable practices in an extraordinarily innovative work and residential environment. Such living labs benefit from highly interactive socio-economic activities among firms, residents, universities and research institutes, as well as governmental institutions and organizations, which all shape the urban innovation system and highlight its role as a bubbling creative cauldron of centers of excellence. This aspect has to be addressed in the context of any port city re-development plan with a lively mix of activities comprising specific patterns, heritage components, demographic developments, economic situations, future potential and international connectivity links, in order to make sustainability work and to improve international competitiveness.

This study has provided, on the basis of structured interviews, an overview of experience and findings that address the socio-economic impacts of the NDSM district in a broader context. In reinventing the port areas, the urban Facebook framework, first developed in Kourtit and Nijkamp [14]) helped us to identify successful strategic policies, and to bring together different expertise to balance: conflicts between the interests and values of a multiplicity of stakeholders; and economic prosperity with social needs and the conservation of eco-systems. In other words, a preference elicitation exercise was organized through the main focus group of users of the NDSM district, while the systematically collected information was analyzed within the urban Facebook evaluation framework, which includes a visual support tool (i.e., the four Urban Faces). This present framework adopts the same general idea, but extends it by recognizing the importance of the visual appearance of the urban cultural ambience and urban future images and ambitions of historic and modern urban districts, based on a stakeholder-oriented (a bottom-up approach). In this framework, visual features and values were integrated in a set of urban future images, which map out different levels of urban planning on the basis of a set of different evaluation criteria. Thus, the framework is a direct action platform that offers social utilities to connect various people who work and live together, supported by high-quality visual assessment tools, for mapping novel redevelopment initiatives.

Taking into consideration each stakeholder's preferences, values, and point of view on the area helps to attract and keep creative minds living and working to develop flourishing, dynamic economies. Each stakeholder has his/her own option and vision for the area. This diversity has helped to create more complex and accurate future images of the area. Thus, a possible (ranging from little to strong) facelift aim of the NDSM-district is to attract and retain creative, high-skilled people, creative firms, etc. to formerly neglected areas in order to achieve sustainable development.

A prerequisite for a promising revitalization policy is that port cities should be able to develop highly innovative strategic approaches to urban planning, conservation and management that really integrate harbor development with urban development. Indeed, both the organizational and economic innovation of the urban space is key to improving the resilience of a port city system, and thus its overall sustainability.

REFERENCES

1. Atkinson, R. Packing the Bunker. Urban Stud. 2006, 43, 819–832.
2. Butler, T. Living in the Bubble. Urban Stud. 2003, 43, 2469–2486.
3. Watt, P. Living in an Oasis. Environ Plan. A 2009, 41, 2874–2892.
4. Kourtit, K.; Möhlmann, J.; Nijkamp, P.; Rouwendal, J. The spatial distribution of creative industries and cultural capital in the Netherlands. J. Cult. Herit. Man. Sustain. Dev. 2013. forthcoming.
5. Rutten, P.; Manshanden, W.; Muskes, J.; Koops, O. De Creatieve Industrie in Amsterdam en de Regio; TNO: Delft, The Netherlands, 2004.
6. SBI. Standaard Bedrijfsindeling (Standard Industrial Classification 1993); Statistics Netherlands (CBS): The Hague, The Netherlands, 1993.
7. Biggs, R.; Raudsepp-Hearne, C.; Atkinson-Palombo, C.; Bohensky, E.; Boyd, E.; Cundill, G.; Fox, H.; Ingram, S.; Kok, K.; Spehar, S.; et al. Linking futures across scales: A dialog on multiscale scenarios. Ecol. Soc. 2007, 12, 17–20.
8. Tress, B.; Tress, G. Scenario visualisation for participatory landscape planning: A study from Denmark. Landsc. Urban Plan. 2003, 64, 161–178.
9. Robinson, J. Future subjunctive: Backcasting as social learning. Futures 2003, 35, 839–856.
10. de Boer, J.; Beekmans, J. Check-in-stedenbouw. Rooilijn 2012, 45, 404–409.
11. Florida, R. Entrepreneurship, Creativity, and Regional Economic Growth. In The Emergence of Entrepreneurship Policy; Hart, D.M., Ed.; Cambridge University Press: Cambridge, UK, 2003; pp. 39–58.
12. Martin, P.; Mayer, T.; Mayneris, F. Spatial Concentration and Firm-Level Productivity in France, Discussion Paper Series 6858, CEPR, Paris, France, 2008.
13. Brain, B.J.L. Cities as systems within systems of cities. Pap. Reg. Sci. Assoc. 1964, 10, 147–163.
14. Kourtit, K.; Nijkamp, P. Strategic choice analysis by expert panels for migration impact assessment. Int. J. Bus. Glob. 2011, 7, 166–194.
15. Nijkamp, P.; Zwetsloot, F.; van der Wal, S. Innovation and growth potentials of European regions: A meta-multicriteria analysis. Eur. Plan. Stud. 2010, 18, 595–611.
16. Nijkamp, P.; Kourtit, K. The 'New Urban Europe': Global challenges and local responses in the urban century. Eur. Plan. Stud. 2012, 1, 1–25.
17. Neuts, B.; Nijkamp, P.; van Leeuwen, E. Crowding externalities from tourist use of urban space. Tour. Econ. 2012, 18, 649–670.
18. Smit, A.J. The influence of district visual quality on location decisions of creative entrepreneurs. J. Am. Plan. Assoc. 2011, 27, 167–184.
19. The term 'creative industries' usually refers to those economic activities that generate both tangible and intangible innovative or knowledge-oriented goods and services, which have an income-generating capacity, while the term 'cultural industries' refers to those activities that have an artistic, historic-social, or entertainment connotation [4].
20. Harvey, D. From managerialism to entrepreneurialism: The transformation in urban governance in late capitalism. Geogr. Ann. Ser. B Hum. Geogr. 1989, 71, 3–17.
21. Nijkamp, P. XXQ factors for sustainable urban development: A systems economics view. Romanian J. Reg. Sci. 2008, 2, 1–34.

Supplemental material is available online at http://www.mdpi.com/2071-1050/5/10/4379/htm.

CHAPTER 10

Between Boundaries: From Commoning and Guerrilla Gardening to Community Land Trust Development in Liverpool

Matthew Thompson

10.1 DILAPIDATED DWELLING

Modernity, it seems, is exemplified not so much by the business park or the airport, but by the dilapidated dwelling (Keiller 2013:54).

Every tenth house or flat seems to be empty and tinned-up. Quite a few have been burned out … The Liverpool Housing Trust has abandoned 20 houses in the area because of persistent vandalism and break-ins. In stark contrast, the successful housing co-ops, whether new build or rehab, stand like oases in a desert of dereliction and run-down blocks of walk-up flats (Towers 1995:230).

Such symptoms of "dilapidated dwelling" reveal a familiar story of post-industrial inner-city decline across the global North. This paper delves into the history and future prospects for regeneration of the particularly deprived neighbourhood of Granby, Liverpool: the specific place described in this scene above. Liverpool's "inner-city problem"—persistent unemployment, deprivation, depopulation, urban shrinkage, housing vacancy, dereliction and

© Thompson, M. (2015), "Between Boundaries: From Commoning and Guerrilla Gardening to Community Land Trust Development in Liverpool" Antipode, 47, 1021–1042. doi: 10.1111/anti.12154. Distributed under the terms and conditions of the Creative Commons Attribution license (http://creativecommons.org/licenses/by/4.0/).

abandonment—has multiple roots and complex contributory factors, not least its economic collapse as a global seaport (Sykes et al. 2013). Conventional large-scale state and market-led regeneration, most recently the Housing Market Renewal (HMR) Pathfinder programme, have largely failed to address these "wicked" problems (Cocks and Couch 2012; Cole 2012). Mutual housing models like the co-ops celebrated above represent a potentially more effective, self-sustaining, and socially just affordable housing tenure and regeneration solution to Liverpool's inner-city problem.

Building on arguments for the re-appropriation of our urban commons and the search for alternatives growing in the cracks of capitalism (Blomley 2004b; Chatterton 2010; Hodkinson 2012a, 2012b; Ward 1985), this paper explores how mutual housing alternatives may be established in disinvested inner-city neighbourhoods, to provide effective institutional blueprints for the democratic stewardship of place. The main part of the paper is an in-depth case study of a campaign in Granby, Liverpool, for a community land trust (CLT) to take back empty homes under community ownership after decades of disinvestment and demolition plans. Incorporated as a legal body in 2011, the "Granby Four Streets" CLT is an innovative attempt to establish an urban CLT as a vehicle for neighbourhood regeneration; making its mark at an opportune moment when large-scale demolition-and-rebuild programmes, notably HMR, have prematurely drawn to a halt following the financial crisis and the imposition of austerity (Pinnegar 2012). After years of anti-demolition campaigning by local residents and failed negotiations between the city council, housing associations, and private developers—a deal has finally been brokered to rehabilitate the four streets as a CLT-led vision.

The CLT vision is for an incremental, self-sustaining, and community-led approach to rehabilitation of housing, public space, and the derelict local high street for new work and retail (Assemble 2013). Redevelopment is envisioned as a piecemeal experiment in community self-help, drawing mostly on local skills and resources, in stark contrast to the speculative development model (Tonkiss 2013). The recent deal with the council gives the CLT ten properties to provide affordable housing for local people in need as well as four corner buildings for community enterprise. Like co-ops, CLTs take land off the market into community ownership, but distinguishing CLTs from other mutual models is the unique capability to separate the ownership of land from the tenure of housing, thereby allowing various interest groups to lease buildings and enabling a partnership approach in the difficult task of redeveloping derelict terraced housing. Granby CLT will lease some houses to its funding and development partner for private rent/sale, as well as to a local eco-housing co-op, the Northern Alliance Housing Cooperative (NAHC), who plan to ecologically retrofit five houses as Terrace 21—"terraced housing for the 21st century"—whilst the land itself remains in CLT ownership for long-term community benefit.

Granby Four Streets CLT is also unique for incorporating the innovative Mutual Home Ownership Society (MHOS) model, which the NAHC co-op intends to use as its legal tenure. Designed to work as a key complementary component of CLTs, the MHOS model has been recently developed by CDS Cooperatives to circumvent the problem of leaseholder enfranchisement that afflicts cooperative tenures (Conaty et al. 2003). The MHOS leases buildings from the CLT, whose constitutional covenants ultimately protect the land from private buy-outs. NAHC were inspired by LILAC (Low Impact Living Affordable Community) in Leeds, the UK's pioneering MHOS development (see Chatterton 2013). LILAC, however, is not coupled with a CLT, so Granby Four Streets treads new ground as the demonstration project of the CLT-MHOS model.

The remainder of Granby Four Streets stock will be transferred to two local housing associations to provide "affordable rent" and shared ownership. Although some activists feel this has diluted the original community vision, the CLT has nonetheless been critically influential in bringing together more powerful development actors around the shared goal of refurbishment for a mixed-tenure neighbourhood. Moreover, the CLT seeks a greater stake in the area than indicated by ownership alone: aspiring towards a "stewardship" role as the over-arching democratic decision-making institution through which all other stake-holders and residents may come together to negotiate and pool resources. This paper explores the challenges of institutionalisation and the promising potential of the CLT model for place stewardship under conditions of austerity and long-term neighbourhood decline.

Originating in the 1960s American civil rights movement to promote black property ownership, CLTs have since been utilised to address the pernicious effects of absentee landlordism, speculative property development and gentrification (DeFilippis 2004). CLTs have mostly been developed for the provision and local collective control of affordable housing, with growing international application (Moore and McKee 2012). But there are real prospects to use the model for neighbourhood regeneration in the UK, following in the footsteps of the US, where CLTs are a relatively well established and growing sector: first institutionalised as a municipal housing programme in Burlington, Vermont in the 1980s (DeFilippis 2004); and in the 1990s by grassroots inner-city community campaigns, notably Cooper Square in New York (Angotti 2007) and Dudley Street in Boston (Medoff and Sklar 1994).

Granby Four Streets is part of an emerging urban CLT movement in the UK, concentrated in London and Liverpool. The first urban CLT campaigns include: the pioneering East London CLT established in 2007 by campaign organisation London Citizens (Conaty and Large 2013); an unsuccessful tenant-led CLT campaign for community ownership of an ex-council estate in Elephant and Castle in London (DeFilippis and North 2004); a failed campaign to acquire empty homes in Little Klondyke, Bootle, just north of Liverpool city centre; and Homebaked CLT in Anfield, Liverpool, a successful arts-led regeneration project for a CLT-owned cooperative bakery and affordable housing funded by Liverpool Biennial (Moore 2014). In contrast to London, the Liverpool campaigns are motivated by the threat of disinvestment and demolition in a shrinking city, rather than the pressures of speculative investment, offering a potentially powerful antidote to problems of capital flight, public disinvestment, and neighbourhood decline. They are among the first attempts to successfully utilise the CLT model as an institutional vehicle for neighbourhood rehabilitation, with an emphasis on collective control of assets that contrasts with the narrower focus on housing affordability of the more established rural CLT movement (Moore and McKee 2012).

The Granby campaign is distinct as a more grassroots initiative, having emerged organically out of resident-led anti-demolition campaigning and activism to reclaim the streets through guerrilla gardening. It shares many characteristics with historic grassroots campaigns against demolition going back to the 1960s, such as Bonnington Square in London and Langrove Street in Liverpool during the 1980s, involving occupations, squatting, and do-it-yourself rehabilitation (Towers 1995)—part of a broader history of self-help housing (Mullins 2010). As a contemporary struggle in this lineage, I hope the Granby case study might shed new light onto these longstanding questions around how legally recognised forms of collective land ownership can be successfully institutionalised out of grassroots activism.

In what follows I explore how the political campaign and formal body of Granby CLT arose from more informal activism and everyday practices of "commoning" (Linebaugh 2014). Although not enough to tackle the severe physical dilapidation, this grassroots activism has nonetheless proved a critical precondition for the CLT's success in attracting vital support and funding to acquire empty homes from the city council. The struggle to build trust with stakeholders has been especially challenging due to a complex local history, but also, I argue, due to the ideological dominance of private property relations within planning practice and property law, which Singer (2000) describes as the "ownership model". Before exploring the case study, I first conceptualise the CLT model in the context of mutual housing, the commons, and the difficulties to institutionalisation posed by the ownership model.

The paper draws on ongoing Economic and Social Research Council (ESRC)-funded doctoral research aiming to understand how radical alternatives to state/market provision of affordable housing have gained traction in the recent history of Liverpool: a city with a particularly rich legacy of mutual housing experiments. The research first identified several pivotal moments of radical experimentation through an extensive desk-based historical study and five scoping interviews with "expert" informants, revealing Granby to be particularly significant in an emerging city-wide CLT movement. From mid-2013 to 2014, I visited Granby and attended the monthly Cairns Street Market; attended community meetings; and conducted 30 in-depth semi-structured interviews with key actors—activists, residents, housing associations and council officers, city politicians, and national policy experts. Interviews were coded for common themes through iterative feedback between conceptual concerns, empirical observation, and broader documentary analysis.

10.2 COMMUNITY LAND TRUSTS: INSTITUTIONAL ARTICULATIONS OF THE COMMONS?

CLTs are one particular model of housing tenure and land ownership within mutualism (Hodkinson 2012b; Rodgers 1999; Rowlands 2009); part of a broader movement for local autonomy and collective ownership of the means of social reproduction (DeFilippis 2004). Mutual housing models provide a third option to the familiar dualist categories of public/private sector, state/market provision—as non-profit, voluntary, community-led, place-based membership associations (Bailey 2012). The key function of mutual models—which range from Garden Cities and tenant co-partnerships, through co-ownership societies, cooperatives, co-housing, mutual homeownership societies, and community self-build—is their capacity to "lock in" the value of land and assets, to protect commonwealth from private expropriation (Conaty and Large 2013). This is where they resonate with the notion of the commons.

In (neo)Marxist thought, the commons stands at the beginning of capitalist history, triggered by initial acts of private enclosure, which formed the basis of primitive accumulation and divorced people from the land and the means of sustaining themselves (De Angelis 2006). This process continues today as accumulation-by-dispossession: the "new urban enclosures" that privatise our "housing commons", those de-commodified dwelling spaces re-appropriated from the market or protected from the full force of exchange relations (Hodkinson 2012a). Commons are constituted by values and practices largely free from transactional market relations: mutual aid, cooperation, solidarity. Commons are simultaneously material resource and social practice, brought into dialectical unity through collective labour, in what Linebaugh (2014) terms acts of "commoning": (inter)active, customary, cooperative social relations rooted in place.

Mutual housing models are imperfect institutional reflections or representations of housing commons. For instance, the socio-material dialectic of the commons is embodied in the CLT form, which describes both the social practices that constitute the organisation and the physical land and assets to be commonly owned. Such models seek to reconnect inhabitants with the means of social reproduction by institutionalising some form of cooperative tenure, or "third estate", in which member tenants cooperatively own land and housing as collective landlords, therefore transcending the landlord–tenant/freehold–leasehold binary that permeates British property law (Rodgers 1999). This mitigates against the inherent alienation and exploitation of the tenant–landlord relation—which Colin Ward (1985) held responsible for the swift physical dilapidation of council housing estates. It does so by providing "dweller control" (Ward 1974): autonomy over the activity of dwelling, which should be seen as a verb as well as a noun, just as the commons is a social activity as well as a material resource. By institutionalising a form of housing commons, mutual housing alternatives have the potential to resolve the deprivation and dispossession at the root of the inner-city problem.

Mutual housing models are necessarily impure pragmatic articulations in legal form of an ideal-type commons, synthesising in complex hybrids different aspects of public, private, and common ownership (Geisler and Daneker 2000). Actually existing commons necessarily entail exclusion as "limited common property": "property held as a commons among the members of a group, but exclusively vis-à-vis the outside world" (Rose 1998:132). Just as their relative autonomy is dependent on external support, internal commoning practices are paradoxically dependent on enclosure from the capitalist outside, thereby threatening to reproduce the social exclusion of private property at a higher scale—a frequent criticism of co-ops. This may be counteracted by the concept of "stewardship", the principle that civil title to land is never absolute, but rather held in trust with duties of care, social responsibility, and accountability in serving the common interests of fellow and future users (Geisler and Daneker 2000). It is morally derived from the idea that property values are only partly "earned" by the labour and investment of the individual owner/occupant, the larger part flowing from what Davis (2010) calls the "unearned social increment": collective value creation emanating from countless contributing actions, transactions, and public investments from local to global.

Stewardship is the ethical principle underpinning the rejection of the individual right to profit in the CLT model (Davis 2010). This unique property regime takes land off the market into local democratic control and, unlike other mutual models, separates the ownership of land from that of buildings, which are leased to members, allowing various housing tenures to co-exist on CLT land. First, this effectively captures the value of land locally—anchoring increasingly mobile capital in place and preventing its extraction—for long-term community benefit and economic security against the threat of financial speculation, public disinvestment or displacement (DeFilippis 2004; Davis 2010); thereby challenging neoliberal financialisation of land by blocking the rights of individuals to profit on their share of equity (Blomley 2004b). Second, this enables "stewardship" of the land for future as well as current inhabitants; overseen by a democratically elected tripartite trust, whose rotating board representatives are equally split between member-residents, expert stakeholders, and the wider community (Davis 2010). The concept of stewardship used here refers specifically to the outward-looking capacity of the CLT model to work for community benefit over mere member-resident benefit, by including broader stakeholder expertise in the democratic management of decommodified land and assets through a trust structure accountable to wider publics; to transcend the exclusivity of ownership through more inclusive access and representation of present, possible, and future user interests of CLT-governed space.

10.3   BETWEEN THE BOUNDARIES OF THE OWNERSHIP MODEL: CHALLENGES FOR INSTITUTIONALISATION

The challenge of institutionalisation of our housing commons is made especially problematic for two reasons. First, articulation of the commons as property rights appears conceptually impossible and politically self-defeating. Private property rights legitimate purely passive individual claims to own and divest of land irrespective of common use, as an abstract deed of entitlement backed up by the state (Singer 2000). Commoning, by contrast, is a horizontal practice with customary rights legitimated autonomously through the very act of their mutual negotiation: a relational claim to shared space justified immanently as an active form of human "doing" (Rose 1994). Articulation as legal rights threatens to codify, ossify, and undermine into passive and alienated relations the highly active, interactive, and organic relations of the commons.

Second, the existing hegemonic system of private property rights—the ownership model—is extremely hostile to other forms of ownership, especially the commons (Singer 2000). The ownership model is the legal foundation of (neo)liberalism, a political discourse and economic project based fundamentally on the institution of private property, rooted in separation and abstraction (Blomley 2004b). It invests absolute control over a clearly delineated space in a single identifiable private owner, whose formal legal title alone bestows entitlement (Singer 2000). It promotes the legal separation of people—between owners/non-owners—and the spatial separation of land, constructing exclusionary walls of capitalist enclosure. By marking territory with visible spatial boundaries, property becomes a "spatialised thing" abstracted from its context, devoid of social relations (Blomley 2004b). This ideological cloaking of property helps make land appear appropriable, transferable, and alienable from its social context. The "right to transfer" and the "right to speculate" in order to profit from property appear as naturalised conditions of land itself, making non-alienable common ownership seem like non-property (Singer 2000). The powerful protection of exchange rights under the ownership model allows the enclosure of urban space into an alienable object, and the extraction of socially produced surplus value. This is the legal DNA of what Lefebvre (2002:305) terms "abstract space": "a naked empty social space stripped bare of symbols"; a globalised net of homogenous quantitative equivalence facilitating exchange relations and erasing the qualitative difference and depth of "lived space".

Neoliberal hegemony is partly maintained by the simplified appeal of the ownership model, whose clear legal "settlement" promises certainty, security, and legibility in otherwise fluid, complex, and contentious social relations (Blomley 2004b). By obscuring the pluralism of property relations and the inherent multiplicity of claims with a neat categorisation of ordering dualisms (Singer 2000)—public/private; owner/non-owner; landlord/tenant—this hides and silences those claims not deemed "proper" forms of (private) "proprietorship" (Rose 1998). Enforcing this divided settlement—between visibility/invisibility; legitimacy/illegitimacy; inclusion/exclusion—is the powerful political vocabulary of property rights: enforceable claims to use or benefit from particular property, sanctioned by the sovereignty of the state. It is only through their translation into legally enforceable property rights that moral common claims gain necessary recognition, protection, and security—an important traverse to be carefully crossed for the long-term survival of collective dweller control.

All efforts to institutionalise mutual housing models must contend with a hostile legal landscape polarised between the public and private realm, and geared towards private homeownership. British property law acknowledges only two types of tenure, inherited from feudalism: freehold and leasehold—landlord/tenant—treating mutual members essentially as either tenants or part-owners (Rodgers 1999). Ironically, leaseholder enfranchisement legislation passed in 1967 to protect tenants from ruthless landlords empowers co-op members to buy out their equity share, thereby threatening the re-imposition of private property relations (Conaty and Large 2013). The co-op movement is lobbying for legislative tenure reform to include a "third estate" (Rodgers 1999): the legal protection required to sustain common property relations over time. Each new mutual model can be seen as the latest historical iteration in institutional vehicles designed to negotiate greater legal protection of the housing commons against enclosure.

Our emerging era of "austerity urbanism" (Peck 2012) has not only compounded the inner-city problem, but also opened up new opportunities for grassroots groups to resist urban enclosure and reclaim space for social reproduction in the interstices of the post-crash city. This is testified by the recent growth and research interest in new forms of grassroots urbanisms, variously prefixed as "guerrilla", "insurgent", "everyday", "do-it-yourself", "interstitial" and "makeshift' (Hou 2010; Iveson 2013; Tonkiss 2013). These practices might include community gardens, occupations, squats, co-ops, and alternative gift economies, and have been characterised as "actually existing commons" (Eizenberg 2012), growing in the "cracks" of the dominant development model (Tonkiss 2013), and pre-figuratively re-imagining urban life as an urban commons (Chatterton 2010). Part of this emphasis on the informal, the temporary, the insurgent, and the micro-scale is no doubt a response to the hegemonic power of the ownership model: the need to form "extra-legal counter-publics" that "operate within legal shadows" to "unsettle" the neoliberal settlement (Blomley 2004b:18). By working silently to reclaim common space between the public/private legal-spatial boundaries in the ownership model, grassroots urbanisms thrive on their invisibility to the system (Iveson 2013). However, by the same token, they are often too informal, ephemeral, disconnected, and localised to properly challenge deeper structural issues to effect lasting urban transformation.

Indeed, the growing literature on alternatives or "alterity"—"the possibility of an economic and political 'other'" (Fuller et al. 2010:4)—highlights the need for some degree of socioeconomic self-sufficiency, or relative autonomy, from mainstream capitalist state structures, through the construction of alternative "circuits of value". Indeed, the long-term success of insurgent attempts to (re)appropriate urban space for control over the means of social reproduction depends on the capacity to exercise collective autonomous control over land and resources (DeFilippis 2004). Paradoxically, under the ownership model, relative local autonomy can only be secured and protected through existing forms of legally sanctioned sovereignty over space, which means actively negotiating and making "deals" with the state and market for access to land and property rights.

Indeed, recent research on self-help housing recognises that the ability to help oneself "from within" is paradoxically dependent on "help from without", from vital external sources of support (Moore and Mullins 2013). Many recent self-help housing initiatives to rehabilitate empty homes for community use have relied on the government's empty homes grants and campaign support from the Empty Homes Agency (Mullins 2010). This is part of the new localism agenda and the UK coalition government's "Big Society"—of neighbourhood planning and community rights to buy/bid/build—in which community asset transfer/acquisition now enjoys cross-party political support (Bailey 2012). In this context, CLTs and self-help housing have received renewed policy interest as part of a growing "third sector" of community-based organisations and social enterprises increasingly turned to by the state to manage assets and deliver public services and regeneration at the neighbourhood scale.

However, British policy interest in the CLT model predates these trends, first imported from the US in the 1990s by British advocates seeking to resolve issues of rural housing affordability, and used by communities in the Scottish Highlands to regain control of assets from quasi-feudal landlords (Moore and McKee 2012). The government-funded National CLT Demonstration Programme from 2006 to 2008 piloted 14 CLT projects (Aird 2009), leading to the formation in 2010 of the National CLT Network, an umbrella organisation that connects and supports member CLTs (National CLT Network 2015). Whilst essential for growth, state support presents the danger of co-optation and dilution of the radical land reform potential and local autonomy of CLTs. The contradictions of institutionalisation, in becoming "state-like", are reflected in the tensions between "scaling up" and "going viral" as alternate forms of replication (Moore and Mullins 2013). Institutionalisation is a delicate balancing act of giving legal and procedural structure to informal grassroots practices without losing the organic social energy and political vision motivating unique projects. The remainder of the paper explores how the power of the ownership model presents complex challenges for the practical institutionalisation of Granby CLT in Liverpool.

10.4   LIVERPOOL: A LABORATORY FOR INNOVATION IN MUTUAL HOUSING EXPERIMENTS?

The Granby campaign is situated in Liverpool's ongoing process of economic and social transformation. From its meteoric rise to world city and leading global seaport in the nineteenth century to its equally dramatic fall from grace following the decimation of its raison d'être, the shipping trade, Liverpool has been an "outrider" of the post-industrial transition, suffering from some of the worst effects of industrial growth and decline, and at the forefront of urban policy innovations to tackle its persistent housing crisis (Nevin 2010; Sykes et al. 2013). Liverpool was the first British city to build public housing in 1869 in response to squalid "back-to-back" tenement housing conditions, later pioneering the UK's first resident-led housing co-ops, and the largest community-led housing trust operating today, the Eldonians (McBane 2008). With the rapid loss of its maritime economic base—capital flight, disinvestment, and unemployment—Liverpool's population halved in under half a century, from a peak of over 800,000 in the 1930s to around 400,000 by 2001 (Cocks and Couch 2012). The inner-city areas of Victorian terraces, once housing thousands of dockers and their families, were disproportionately hit by the decline, with severe depopulation, dereliction, and deprivation: by the 1990s some of these neighbourhoods had vacancy rates of over 30% (Nevin 2010).

The post-war municipal policy response to poor housing conditions was large-scale demolition, or "slum clearance" programmes with around 160,000 inner-city residents decanted to new towns and estates on the metropolitan periphery (Sykes et al. 2013). This exacerbated inner-city decline by removing working populations from economically fragile areas, thereby designing-in-dereliction. At the epicentre of these clearances is Granby, a particularly deprived inner-city ward in the south-central postcode of Liverpool 8, renowned for its rich cultural history, ethnic diversity, and faded architectural grandeur (Merrifield 2002). Granby is home to one of the UK's oldest black communities—a long and complicated history entwined with place that reaches back to Liverpool's roots in Atlantic trade—and witnessed one of its most virulent and violently repressed riots in living memory, against poverty, institutional racism, and police brutality (Frost and Phillips 2011). Not only did the "1981 Uprising" imprint the area with a perceived social stigma—thereby reinforcing decline—it also created mutual mistrust between city authorities and local residents, some of whom believe the council has engaged in a deliberate programme of managed decline (resident interviews 2014).

Yet Granby's decline has provoked community resistance and social innovation through mutual alternatives. The Shelter Neighbourhood Action Programme (SNAP), the pioneering action research project run by the homelessness campaign organisation Shelter from 1969 to 1972, was set up to resolve Granby's endemic deprivation and appalling "slum" housing conditions (McConaghy 1972). SNAP helped establish the country's first rehabilitation housing cooperatives, in turn inspiring a flourishing new-build housing co-op movement in the 1970s—leaving a legacy of over 50 co-ops across Liverpool (Lusk 1998). This was motivated by widespread agitation for better housing conditions among residents living in insanitary and poorly maintained terraces and tenements; driven by resistance to displacement and community fragmentation (interviews 2013). Colin Ward's (1974) radical ideas for "dweller control" were influential in the development of the Weller Streets in Granby, the UK's first truly resident-led fully mutual new-build co-op (McDonald 1986). The subsequent rhizomatic spread of new build co-ops across Merseyside was deeply rooted in an innovative programme of tenant education in cooperative principles, architectural design, and housing development regulations (interviews 2013). It was spearheaded by the secondary co-op development organisation, CDS, working with local architectural firms to innovate participatory design methods that enabled working class residents to design, develop, and manage their own homes in an unprecedented process of collective dweller control.

The exceptionally generous funding regime and supportive infrastructure of this period facilitated the growth of co-ops as well as housing associations, which have since expanded to become the most powerful property development players in Granby today (Lusk 1998). Indeed, Liverpool's large professionalised housing associations started out as small non-profit charitable trusts and co-op agencies. The largest association operating in Granby today, Plus Dane, is the direct heir of CDS, which it absorbed in the 1990s. Subsequent political opposition during the 1980s from the Militant-dominated Labour council threatened the co-op movement with "municipalisation" and extinction, yet also galvanised other community groups, such as the Eldonians, into action (Frost and North 2013). Neoliberal reforms have since put an end to co-op development, reflecting broader trends towards the privatisation of public housing, through Right to Buy and stock transfer of council housing to an increasingly market-led housing association sector (Ginsburg 2005). Whilst the co-op movement has been constrained from further development by neoliberal policies it has none-theless opened up the political space between public and private to think cre-atively about how to resuscitate a problematic area like Granby.

Granby's ageing pre-1919 housing stock has long passed its planned physi-cal lifecycle—despite council-funded refurbishments—and worsening socio-economic conditions have conspired to create a downward spiral of decline and dilapidation (Merrifield 2002). Post-war planning mistakes contributed to this decline: redirecting and building over the top end of the once-bustling neighbourhood shopping avenue, Granby Street, thereby severing Granby from its vital connection with the city centre as an arterial through-flow for urban activity and consumption (housing officer interview 2013). Further council-led demolition-and-rebuild programmes attempted to tackle the dereliction, replacing most terraces with lower density estates, leaving only four original streets, known as the "Granby Triangle". These four streets map neatly onto the original SNAP boundaries, suggesting that early rehabilitation efforts have been relatively successful. From the 1990s, the council began buying up hous-ing association properties—emptying them of their tenants—and offering mar-ket prices to the small minority of remaining owner-occupiers. A vocal group of homeowners, organised as Granby Residents' Association, refused to move and campaigned to save the streets. Described by an ex-council officer as the "final battleground" (interview 2013), these four streets became centre-stage to a bitter conflict fought between the council and the small minority of remaining residents. The resistance attracted the support of national lobby organisations Empty Homes Agency and SAVE Britain's Heritage, helping raise the media pro-file of the campaign to rehabilitate rather than demolish empty terraces.

Conflict intensified with the commencement of HMR Pathfinders, the con-troversial £2.3 billion national programme rolled out from 2003 to 2011 across nine de-industrialised northern English inner-cities, notably Liverpool, whose city council helped pioneer and lobby for government funding (Cole 2012; Nevin 2010). HMR Pathfinders aimed for long-term structural change in fail-ing housing markets through part-refurbishment and large-scale demolition of "obsolete" Victorian terraces and replacement with a more "sustainable" mix of tenures (Webb 2010). Part of the mixed communities agenda, HMR has been critiqued as state-led gentrification, remaking place in the image of a new target middle class population (Allen 2008), and for conceiving lived neighbourhoods as abstract sub-regional markets, conceptualising the "city-as-property" over the "city-as-inhabited" (Pinnegar 2012). From a Lefebvrean perspective (Wilson 2013), HMR represents the domination of "abstract space", based on exchange value, over the use values of "lived space".

Liverpool's HMR Pathfinder, "New Heartlands", earmarked around 70,000 houses in an inner-city ring for demolition/refurbishment, initially forecasting £3 billion public/private investment until planned completion in 2018 (Nevin 2010). Liverpool was divided into four "Zones of Opportunity"—or "ZOOs"— each appointed a single preferred developer to work in partnership with the area's leading housing association, and accountable to a governing board of stakehold- ers, which, unlike previous regeneration programmes, included no local resident representation (Cole 2012). In a tragic repeat of history, ZOOs mapped closely onto the 1960s slum clearance areas: a landscape witness to more than two gen- erations of regeneration (Sykes et al. 2013). This relentless focus on one mono- lithic solution to complex neighbourhood contexts—with little opportunity for

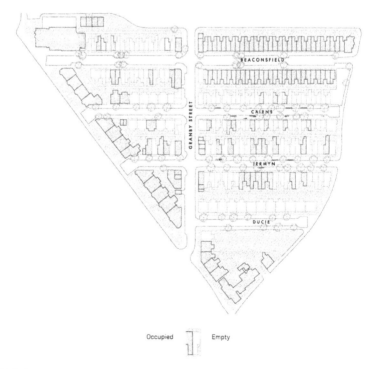

FIGURE 10.1   Map showing vacancies in Granby Four Streets (source: Assemble 2013; reproduced here with permission)

piecemeal community projects—demonstrates the enduring influence of the ownership model over regeneration thinking in Liverpool.

The failure of HMR to resolve the inner-city problem—at least in part attributable to the premature withdrawal of state funding mid-way through its planned lifecycle in 2011 in the context of post-crash austerity—is now all too evident in the swathes of vacant land and empty tinned-up properties across HMR clearance zones. In Granby, HMR did fund significant refurbishment but most of the area was left to crumble into dereliction, still in council ownership but without funds for either demolition or refurbishment. Today, there are 128 vacant boarded-up houses and shops, leaving only around 60 households still lived in (see Figure 1).

10.5 GROWING GRANBY FROM THE GRASSROOTS

Long before the withdrawal of HMR, the remaining residents in Granby had already begun to resist its adverse effects—properties boarded up, streets collecting rubbish, attracting vandals, houses literally falling in—by cleaning pavements, clearing rubbish, and reclaiming the derelict streetscape as a community garden. They placed potted plants and garden furniture out on pavements, painted derelict house frontages with murals, and grew plants and flowers up buildings (see Figure 2). Much of this was preceded by a council-funded adult education programme on ecology and gardening called "Growing Granby", which entrusted a nearby vacant plot to local residents via a short-term lease by housing association Liverpool Mutual Homes for a community garden, as well as inspiring more radical ideas for a "DIY People Plan" reimagining Granby as a "backyard commons" (Grant 2011). Yet the insurgent acts of guerrilla gardening that have transformed the Granby Triangle into what is known as the "Green Triangle" sprang forth more spontaneously from residents themselves—distinguished from "Growing Granby" "because we work in public space, not behind railings on private land" (activist interview 2014).

Working without permission from the council, these guerrilla gardeners engage in everyday acts of "commoning": bringing the domestic, intimate spaces of their homes out into the public streetscape, sharing it with others, and creating a distinctive hybrid community garden that mixes domesticity, privacy, communality, and public openness, bearing the hallmarks of an "actually existing commons" (Eizenberg 2012). This blurs the boundaries of public and private space, representing what Blomley (2004b:15) calls "creative acts of resistant

FIGURE 10.2   Green Triangle guerrilla gardening (source: photos by author)

remapping" of the official abstract map of the ownership model. In blurring these boundaries such insurgent acts are informal and unrecognised forms of ownership—an "imagined proprietorship" (Blomley 2004a) or an "un-real estate" (Rose 1994), highlighting the organic and active aspects of ownership as a process of human "doing" (Rose 1994). This is a stark refutation of the ownership model and its insistence that only two moments of action matter: acquisition and transfer (Blomley 2004a). Green Triangle commoning also cuts across political and social distinctions among residents, who have forged common bonds despite diverse worldviews through communal cleaning, planting, and tending (activist interview 2014). However, these practices are largely confined to a small number of remaining homeowners, highlighting how "commoning is exclusive inasmuch as it requires participation. It must be entered into ... This is why we speak neither of rights nor obligations separately" (Linebaugh 2014:15). There is a need for the Granby CLT to seek greater inclusion of wider publics and more direct participation of other residents for democratic legitimacy.

One way progress is being made in this direction is through the monthly street market, which, since its inception off the springboard of guerrilla gardening, has become a symbol of resistance and community hub for small-scale economic and cultural activity—a local legend, attracting over 200 people over a day from all over Liverpool and beyond (see Figure 3). Once a month, the local community comes together in celebration—setting up stalls selling everything from everyday essentials to artwork—with live performances from local musicians, diverse cuisine cooked on-site, and dancing amongst the medley of shoppers, sellers, and wanderers. This do-it-yourself experiment is a tentative move towards constructing a relatively autonomous "circuit of value" (Fuller et al. 2010), with plans to acquire four corner buildings on Granby Street as community-owned enterprises, studios, cafes, and shops as part of its regeneration into the bustling shopping avenue it once was.

FIGURE 10.3   Granby Four Streets market (source: photos by author)

These do-it-yourself developments have both progressive and regressive potential, containing the contradictions of what Tonkiss (2013) calls "interstitial urbanism". Their creative and pioneering endeavour to take back streets left to decay by austerity politics is both a crack in the ownership model, prefiguring an actually existing commons, and simultaneously an unwitting agent of austerity urbanism, taking up the slack in the paralysed development model and filling the gap left by the retreating state to productively reuse derelict housing when all else has failed. Granby's Green Triangle thus fulfils an ambiguous double-role, vis-à-vis "roll-with-it" neoliberalism (Keil 2009)—the normalisation in everyday life of entrepreneurialism, creativity, self-reliance, flexibility, and do-it-yourself initiative as a means to facilitate capital accumulation. Green Triangle activists, mostly women associated with the city's artistic milieu, enact a certain bohemian habitus which may act to alienate or exclude other social groups from the area, and which plays into "creative class" politics and city branding, potentially planting the seeds for green gentrification as Liverpool's economy recovers.

Indeed, this has attracted the interest of other creative types in the area. The Northern Alliance Housing Cooperative (NAHC)—a small group of idealistic young professionals, designers, and postgraduate students living locally and looking for empty homes to retrofit into mutualised eco-homes—was established in direct response to the Green Triangle. NAHC's founder was originally inspired by the creative endeavour evident in the "beautiful" transformation of the four streets to explore the idea of a co-op and present it to residents (NAHC interview 2013). Likewise, the ex-housing officer who became Granby's community organiser, helping channel divergent creative energies into a common vision, offered his services after first being seduced by the green activism. Perhaps the most vital support came from a private social finance company, HD Social Investments (HDSI), personally backed by what CLT members describe as "the mystery millionaire" (activist interviews 2013). This former stockbroker from Jersey had sent his researcher out around the country to search for a socially worthwhile project in which to invest finance capital for a small return—described by a CLT activist as "philanthropy at 5%"—and came across Granby through auspicious links with SAVE Britain's Heritage. Piquing his interest in Granby was not just the Victorian architectural assets found right across inner-city Liverpool, but the proactive do-it-yourself ethos and social entrepreneurialism of Granby residents breathing life back into their faded grandeur.

Conflicts of interest between the community and the private investor may well play out in due course, but so far HDSI have provided crucial financial support: considerable low-interest loans as well as the funding and expertise required to successfully apply for several grants, such as Nationwide Foundation and government's Empty Homes funds, each worth £125,000 (interviews 2014). Working with CLT members, HDSI has also commissioned a persuasive design statement from the innovative London-based architecture collective, Assemble, which sets out a practical plan to acquire and refurbish 27 of the 128 vacant, boarded-up empty homes in the four streets as a mix of affordable homes, as part of a long-term vision to rehabilitate the other empty homes and revive the neighbourhood's economic backbone, Granby Street (Assemble 2013). Under the creative direction of Assemble, the CLT is working with HDSI and NAHC as joint partners to realise this vision—each hoping to take on properties and manage them as different tenures—but with the CLT as the ultimate umbrella institution under which all other partners and legal ownership of the land are organised.

A large part of the broad community mandate for the CLT model is its capacity to incorporate the co-op and other tenure types, integrating divergent property interests, and the democratic trust governance structure, enabling wider stakeholder participation for long-term place stewardship for community benefit over resident-member benefit. CLT membership extends throughout the L8 postal district, beyond the immediate Granby Triangle, and so the CLT recognises its scalar contributory relationship with surrounding urban areas. Members meet regularly to discuss CLT affairs and democratically elect representatives onto the trust management board, whose membership of 12 periodically rotates, with tripartite representation of member residents, the wider local community, and key stakeholders. The latter third includes representatives from Plus Dane and Liverpool Mutual Homes, the council, as well as crucial financial and technical expertise in development. The diverse black community are actively engaged as stakeholders: the Men and Women's Somali Groups each have board representation, as does the Steve Biko housing association, established in 1982 to provide local black community access to social housing in the context of racial discrimination, and now helping develop and deliver the CLT housing allocations policy. Tenants displaced by HMR are represented in the wider community third, to be afforded a "right to return" in CLT housing allocations; but it remains unclear how the very limited number of houses will be fairly distributed among the much larger number of evicted tenants.

Indeed, such apparent inclusivity is not without internal tensions: the CLT is marked by what many describe as tense politics. The local black community has a long historical attachment to Granby, which, coupled with perceived injustices of persecution, produces a strong sense of place entitlement. Emerging conflicts between longstanding resident homeowners and NAHC newcomers, who have nonetheless lived in the surrounding area for many years, reflect opposing ethical perspectives on rights to place: personal historical attachments to place versus productive contribution through active improvement. NAHC members bring professional skills in ecology, architecture, and planning to the campaign process—critical in persuading the council to even consider the CLT idea—and claim inclusion on the basis of their innovative project to retro-fit five of the empty homes into cutting-edge eco-houses to be managed cooperatively through a MHOS. These claims to expertise, however, may also act to exclude, and efforts need to be made to engage other residents in a more mutual and open learning process.

10.6   TRUST: THE CLUE'S IN THE TITLE!

Gaining the support of the council, as the primary gatekeeper, is essential for successful acquisition. From the council perspective, the burden of proof lies firmly on the CLT to demonstrate its social responsibility to manage assets, and to convince local government of the merits of transferring a large quantity of public assets to an untested community-owned organisation. A local architect/NAHC/CLT activist states the problem:

> We have to prove that we can do something before people trust, because that issue of trust goes both ways as well … local residents don't trust the city council, the city council don't trust local residents to do anything other than kick up a fuss … Hopefully that would get easier … breaking down the barriers that have been built up over the last ten years with HMR … and a certain fear at the council level … just trusting people to do the best for the neighbourhood doesn't really seem to be there. I think it's there now with some of the members but it's still not there with all of the officers; that's an institutional culture thing, which I expect takes decades to change.

Trust is the magic ingredient holding the entire CLT endeavour together. HMR in Granby stands at the end of a long complicated history of mutual mistrust between council and community, first flaring in the 1981 Uprising and now threatening to paralyse collaborative decision-making over the future of the area. Residents feel a powerful sense of resentment and injustice that their homes and community have been "stolen" from them by the council (resident interviews 2013)—an oppositional position posing additional barriers to negotiating a mutually satisfactory solution.

The absence of trust is evident in the council's decision, in the wake of HMR's cancellation, to tender the four streets for "best value" bids, entering into year-long negotiations with a private development company, Leader One, whilst CLT ideas were side-lined. Such a competitive logic—pitting parties against each other—is a manifestation of the ownership model, formalised in the 1980s by compulsory competitive tendering policies (Hodkinson 2011). This winner-takes-all approach is attractive to councils who can settle a definite contract with one single responsible and liable owner, but which imposes severe entry barriers for smaller community-led projects without the resources or expertise of private companies. It is also risky: the Leader One deal collapsed under unreasonable demands for the council to underwrite any losses, the admirable refusal to effectively privatise profit and socialise risk.

During this process, activists approached Leader One to propose a partnership, which the company briefly entertained. An NAHC founder tells of how it was Leader One, during negotiations with the council, who first suggested to him that "the council isn't interested in having a cooperative there", explaining that "if you can make it like some kind of ownership thing, then we might be a bit more interested"; persuading NAHC to pursue MHOS as a more palatable mutual solution than a conventional co-op. The preference for mutual homeownership over a traditional co-op is as much about the perceived fear and mistrust of common property regimes that sit outside the familiar categories of the ownership model—assuaged by the semantic association with individual "homeownership"—as it is with the actual workings of the MHOS model itself, which are more akin to a co-op than its name suggests (Chatterton 2013). This is where its power lies in playing the language game of private property rights, and potentially using this brand advantage as a way to leverage support from otherwise sceptical gatekeepers.

It was only the austerity-driven failure of Leader One that eventually turned attention towards the CLT vision: the only viable option left on the table. A change in council mind-set was already evident in its self-help "homesteading" plan. Empty Homes funding has been made available to sell empties for £1 to individuals with local connections to restore through do-it-yourself labour on the proviso that certain conditions are met, such as living in the house for at least five years without sub-letting (Crookes and Greenhalgh 2013). Such a piece-meal approach is perhaps too individualised to effectively tackle a large area of empties, having only been tested with a handful of properties in Granby. Yet it signals a break with the dominant speculative development model.

CLT partners have taken inspiration from this self-help method to come up with "community homesteading" (activist interview 2014). They plan to develop CLT houses on a one-by-one basis, drawing on the do-it-yourself self-build techniques of homesteading but employing resources and labour from across the entire community. They hope to establish relationships with local colleges to help train young people in craft and construction in return for lower labour costs, thereby strengthening financial viability and embedding development in the local economy. This disrupts both the spatial and temporal logic of the neoliberal urban development process: the "sharp-in/sharp-out" model, which "assumes a division between the makers and the users of space" in the fallacy of the "end-user" (Tonkiss 2013:320), and alienates existing residents from the process, whose lives are put on hold or displaced entirely. Community homesteading, by contrast, transcends this division through a more socially participatory, temporally incremental, albeit spatially piecemeal approach towards securing collective dweller control.

10.7   CONCLUSION: THE CONTRADICTIONS OF INSTITUTIONALISATION

This paper has advanced two opposing imaginaries of housing ownership and neighbourhood regeneration. The first describes the dominant ownership model, which sees ownership and dwelling as externally related, with property appearing separable from its social context, enabling abstraction and exchange on the global market, as a form of abstract space. The alternative is a utopian imaginary of internal relations, in which the social and material aspects of dwelling are dialectically entwined. Whereas the former is founded on a disconnection between producer/consumer—the alienation of landlord/tenant, owner/occupier—the latter reconnects maker with user, developer with dweller, through collective dweller control. Active doing is emphasized over passive entitlement. This perspective materialises as a collective self-help regeneration method, drawing on do-it-yourself techniques and practices of commoning—tentatively expressed in Granby's guerrilla gardening and community homesteading. By virtue of the self-securing nature of British, or Anglo-Saxon, private property rights, common ownership must be actively and creatively claimed, through unconventional insurgent tactics that work beyond the law. Granby's grassroots practices are essentially "imagined"—but politically powerful—claims for a common right to place. Without licence, residents have acted upon public space as if it were their own: actively resisting managed decline to "take ownership" and reclaim lived space from the abstract space of HMR. However, the long-term survival and viability of collective control over the means of social reproduction is paradoxically dependent on state support to authorise and finance community acquisition of land and recognise its legal ownership.

Actually existing commons are neither free from contradictions nor immune to human power relations. They construct their own walls within—and boundaries without—as necessarily exclusive enclosures that protect against more pernicious enclosures. Mutual housing models are essentially pragmatic compromises made with a hostile legal landscape that attempt to express mutual relations in institutional form. As forms of housing, they are complex hybrid social spaces, combining the necessary privacy of the home with more cooperative social relations for the democratic governance of land. The great strength of the CLT model is its flexibility in the face of hegemony: its incorporation of multiple tenures enabling diverse interest groups, stakeholders, and sources of support to govern together through trust. This emphasis on stewardship over ownership—community benefit over mere member benefit—is a promising avenue towards overcoming the inherently exclusionary dynamics of housing commons. The political potential of stewardship to transcend the gap between common ownership and public trust lies in this capacity of the CLT structure to incorporate wider publics in democratic decision-making; acting as an outward-looking counterbalance to the necessarily inward-looking closure of housing commons. Further empirical research is required as Granby CLT develops to assess this potential and investigate the actual effects of governance practices, particularly housing allocation decisions, on distributive justice and social relations.

The Granby campaign is a novel experiment in the CLT-MHOS model and community homesteading; but to be more than just an isolated one-off experiment, the issue of replication is fundamental. Granby's success so far appears to stem from contextual particularities peculiar to time and place: chance encounters with co-op activists, community organisers, and social investors; the unique local history of collective action and innovation in cooperative housing; and the window of opportunity opened up by austerity urbanism. It was only through the moratorium placed on monolithic demolition-and-rebuild schemes that the CLT became attractive to Liverpool policymakers—a last-ditch option when all other standard approaches had been exhausted. Austerity also demands funding from sources other than the state, in this case from the HDSI "mystery millionaire". Such an emerging role for social investment and the reliance on private capital raises many questions over the accountability, viability, and replicability of such schemes, perhaps made too vulnerable to the whims of philanthropic capital. However, by understanding the socio-political dynamics of groundbreaking projects first tested out under such extreme conditions, I hope to have revealed insights for the political potential of mutual housing projects in other contexts, with similar catalytic conditions.

The myriad preconditions for urban CLT campaigns to re-appropriate empty homes for community use exist in countless other places, and we can see seeds taking root in similar ex-HMR inner-city contexts, for instance in Middlesbrough (MCLT 2015). Lessons can be learnt from an unsuccessful campaign in Little Klondyke, Bootle, just north of Liverpool, which, despite sharing many characteristics with Granby—deprived ex-HMR inner-city neighbourhood of derelict terraced housing whose residents fought a bitter battle against demolition—nonetheless failed to gain the vital consent of the local authority to sign off on an otherwise successful grant application for some £5 million from DCLG's Empty Homes Community programme, secured with the help of the National CLT Network, the Empty Homes Agency, and SAVE Britain's Heritage (activist interview 2013). Sefton Council's refusal to support the funding programme may indicate entrenched ideological beliefs in the ownership model, but may also reside in the lack of local participation in the campaign, struggling to find the minimum 12 residents required to constitute a functional CLT board. Such a contrasting story shows the essential ingredient in Granby's success to be the dynamic and creative grassroots activity that first spurred others to seriously consider the merits of the CLT. It was only through this performative demonstration to city authorities and potential allies of a local collective will to take on the stewardship of a disinvested space that vital funding streams and development expertise were ever secured. A fundamental barrier is therefore the considerable burden on local volunteering energies—residents' proactive capacities, skills, and motivations to engage in complicated campaign and development processes—and their deeply problematic uneven spatial distribution; raising serious concerns for the viability and systematic replication of such projects, especially in the poorest neighbourhoods where they are most needed.

The challenge of replication and institutionalisation hinges on this tension: between, on the one hand, inspiring, mobilising, and sustaining the intense political campaign energy and grassroots practices of commoning that are the lifeblood of common ownership institutions; and, on the other, the need for legal definition, professional expertise, and scaling up into institutional structures. If such mutual experiments are to take root and grow in other disinvested contexts, more systematic support and coordination from intermediary bodies, such as regional-scale umbrellas, is required to nurture the seeds and plant new seeds through viral transfer: to bring together localised experiments into a more connected movement; to enable mutual learning, knowledge sharing, and resource pooling, whilst avoiding the pitfalls of professionalisation. In a promising move, the National CLT Network (2015) has recently secured social investment funds for an Urban CLT Project to offer £10,000 grants to 20 demonstration projects to specifically support the difficult transition stages after start-up, such as negotiating with land acquisition.

Insights may be drawn from Liverpool's housing history: progressive lessons from the cooperative education and design democracy at the heart of the 1970s new build co-op movement; and also warning signs. Just as the city's huge housing associations, recently helping deliver HMR demolition, started out as place-based charitable trusts—CDS morphing into Plus Dane, for instance—so too is there a danger that Granby, like other CLTs, might eventually mutate into an unwieldy concern with large-scale property interests and little connection to people or place. A key question for further research is how to secure lasting collective dweller control without becoming just another part of the shadow state, overloaded with unwanted public service delivery responsibilities. In seeking to develop and replicate successful common ownership institutions, we run the risk of diluting, paralysing, and fossilising into inflexible bureaucratic structures the informal, spontaneous, and creative energies of commoning which animate radical collective action.

REFERENCES

1. Aird J (2009) Lessons from the First 150 Homes: Evaluation of the National Community Land Trust Demonstration Programme, 2006–2008. Salford: Community Finance Solutions
2. Allen C (2008) Housing Market Renewal and Social Class. London: Routledge
3. Angotti T (2007) New York for Sale: Community Planning Confronts Global Real Estate. Cambridge: MIT Press
4. Assemble (2013) The Granby Four Streets. Liverpool: HD Social Investments
5. Bailey N (2012) The role, organisation, and contribution of community enterprise to urban regeneration policy in the UK. Progress in Planning 77(1):1–35
6. Blomley N (2004a) Un-real estate: Proprietary space and public gardening. Antipode 36(4):614–641
7. Blomley N (2004b) Unsettling the City. London: Routledge
8. Chatterton P (2010) Seeking the urban common: Furthering the debate on spatial justice. City 14(6):625–628
9. Chatterton P (2013) Towards an agenda for post-carbon cities: Lessons from Lilac, the UK's first ecological, affordable cohousing community. International Journal for Urban and Regional Research 37(5):1654–1674
10. Cocks M and Couch C (2012) The governance of a shrinking city: Housing renewal in the Liverpool conurbation, UK. International Planning Studies 17(3):277–301
11. Cole I (2012) Housing market renewal and demolition in England in the 2000s: The governance of "wicked problems". International Journal of Housing Policy 12(3):347–366
12. Conaty P, Birchall J, Bendle S and Foggitt R (2003) Common Ground—for Mutual Home Ownership: Community Land Trusts and Shared-equity Cooperatives to Secure Permanently Affordable Homes for Key Workers. London: New Economics Foundation and CDS Co-operatives
13. Conaty P and Large M (eds) (2013) Commons Sense: Co-operative Place Making and the Capturing of Land Value for 21st Century Garden Cities. Manchester: Co-operatives UK
14. Crookes L and Greenhalgh W (2013) DIY Regeneration? Turning Empty Houses into Homes Through Homesteading. Sheffield: University of Sheffield and Empty Homes

15. Davis J E (2010) The Community Land Trust Reader. Cambridge: Lincoln Institute of Land Policy

16. De Angelis M (2006) The Beginning of History: Value Struggles and Global Capital. London: Pluto

17. DeFilippis J (2004) Unmaking Goliath: Community Control in the Face of Global Capital. London: Routledge

18. DeFilippis J and North P (2004) The emancipatory community? Place, politics, and collective action in cities. In L Lees (ed) The Emancipatory City? Paradoxes and Possibilities (pp 72–88). London: Sage

19. Eizenberg E (2012) Actually existing commons: Three moments of space of community gardens in New York City. Antipode 44(3):764–782

20. Frost D and North P (2013) Militant Liverpool: A City on the Edge. Liverpool: Liverpool University Press

21. Frost D and Phillips R (2011) Liverpool '81: Remembering the Riots. Liverpool: Liverpool University Press

22. Fuller D, Jonas A E G and Lee R (eds) (2010) Interrogating Alterity: Alternative Economic and Political Spaces. Farnham: Ashgate

23. Geisler C and Daneker G (2000) Property and Values: Alternatives to Public and Private Ownership. Washington, DC: Island Press

24. Ginsburg N (2005) The privatization of council housing. Critical Social Policy 25(1):115–135

25. Grant J (2011) Backyard commons. In R MacDonald (ed) DIY City (pp 51–59). Liverpool: Liverpool University Press

26. Hodkinson S (2011) Housing regeneration and the private finance initiative in England: Unstitching the neoliberal urban straitjacket. Antipode 43(2):358–383

27. Hodkinson S (2012a) The new urban enclosures. City 16(5):500–518

28. Hodkinson S (2012b) The return of the housing question. ephemera 12(4):423–444

29. Hou J (ed) (2010) Insurgent Public Space: Guerrilla Urbanism and the Remaking of Contemporary Cities. London: Routledge

30. Iveson K (2013) Cities within the city: Do-it-yourself urbanism and the right to the city. International Journal of Urban and Regional Research 37(3):941–956

31. Keil R (2009) The urban politics of roll-with-it neoliberalization. City 13(2/3):230–245

32. Keiller P (2013) The dilapidated dwelling. In id. The View from the Train: Cities and Other Landscapes (pp 51–63). New York: Verso

33. Lefebvre H (2002 [1961]) Critique of Everyday Life, Vol. II. New York: Verso

34. Linebaugh P (2014) Stop, Thief! The Commons, Enclosures, and Resistance. Oakland: PM Press

35. Lusk P (1998) Citizenship and Consumption in the Development of Social Rights: The Liverpool New-build Housing Co-operative Movement. Unpublished thesis, University of Salford

36. McBane J (2008) The Rebirth of Liverpool: The Eldonian Way. Liverpool: Liverpool University Press

37. McConaghy D (1972) Another Chance for Cities: SNAP 69–72. Liverpool: Shelter Neighbourhood Action Project

38. McDonald A (1986) The Weller Way: The Story of the Weller Street Housing Cooperative. London: Faber and Faber

39. MCLT (2015) Middlesbrough Community Land Trust. http://www.middlesbroughclt.org.uk/ (last accessed 12 January 2015)
40. Medoff P and Sklar H (1994) Streets of Hope: The Fall and Rise of an Urban Neighborhood. Boston: South End Press
41. Merrifield A (2002) Dialectical Urbanism. New York: Monthly Review Press
42. Moore T (2014) Affordable Homes for Local Communities: The Effects and Prospects of Community Land Trusts in England. University of St Andrews Centre for Housing Research: St Andrews
43. Moore T and McKee K (2012) Empowering local communities? An international review of community land trusts. Housing Studies 27(2):280–290
44. Moore T and Mullins D (2013) "Scaling-Up or Going-Viral: Comparing Self-Help Housing and Community Land Trust Facilitation." Third Sector Research Centre Working Paper No. 94. http://www.birmingham.ac.uk/generic/tsrc/documents/tsrc/working-papers/working-paper-94.pdf (last accessed 12 January 2015)
45. Mullins D (2010) "Self-Help Housing: Could it Play a Greater Role?" Third Sector Research Centre Working Paper No. 11. http://www.birmingham.ac.uk/generic/tsrc/documents/tsrc/working-papers/working-paper-11.pdf (last accessed 12 January 2015)
46. National CLT Network (2015) National Community Land Trust Network. http://www.communitylandtrusts.org.uk/home (last accessed 12 January 2015)
47. Nevin B (2010) Housing market renewal in Liverpool: Locating the gentrification debate in history, context, and evidence. Housing Studies 25(5):715–733
48. Peck J (2012) Austerity urbanism: American cities under extreme economy. City 16(6): 626–655
49. Pinnegar S (2012) For the city? The difficult spaces of market restructuring policy for the city? International Journal of Housing Policy 12(3):281–297
50. Rodgers D (1999) New Mutualism: The Third Estate. London: The Co-operative Party
51. Rose C M (1994) Property and Persuasion: Essays on the History, Theory, and Rhetoric of Ownership. Boulder: Westview Press
52. Rose C M (1998) The several futures of property: Of cyberspace and folk tales, emission trades and ecosystems. Minnesota Law Review 83:129–182
53. Rowlands R (2009) "Forging Mutual Futures—Co-operative and Mutual Housing in Practice: History and Potential." Phase 1 Research Report to the Commission on Co-operative and Mutual Housing, Centre for Urban and Regional Studies, University of Birmingham
54. Singer J W (2000) Entitlement: The Paradoxes of Property. New Haven: Yale University Press
55. Sykes O, Brown J, Cocks M, Shaw D and Couch C (2013) A City Profile of Liverpool. Cities 35:299–318
56. Tonkiss F (2013) Austerity urbanism and the makeshift city. City 17(3):312–324
57. Towers G (1995) Building Democracy: Community Architecture in the Inner Cities. London: Routledge
58. Ward C (1974) Tenants Take Over. London: Architectural Press
59. Ward C (1985) When We Build Again: Let's Have Housing That Works! London: Pluto
60. Webb D (2010) Rethinking the role of markets in urban renewal: The housing market renewal initiative in England. Housing, Theory, and Society 27(4):313–331
61. Wilson J (2013) "The devastating conquest of the lived by the conceived": The concept of abstract space in the work of Henri Lefebvre. Space and Culture 16(3):364–380

The Sustainable and Healthy Communities Research Program: The Environmental Protection Agency's Research Approach to Assisting Community Decision-Making

Kevin Summers, Melissa Mccullough, Elizabeth Smith, Maureen Gwinn, Fran Kremer, Mya Sjogren, Andrew Geller, and Michael Slimak

11.1 BACKGROUND

John Muir, "When we try to pick out anything by itself, we find it hitched to everything else in the Universe" [1].

Hundreds, perhaps, thousands, of times a day, communities throughout the United States (U.S.) make decisions on infrastructure, schools, roads, facilities, and a host of other issues. In most of these cases, the decisions are clearly made in good faith and with an outcome in mind—often a single short-term outcome targeted at the primary basis for the decision [2]. By this statement, we are not intending to be condescending to local decision makers (indeed similar arguments could be forwarded for decision making at the state and federal levels) but rather are simply stating an observation that many decisions result in unintended consequences. These consequences are often due to lack of holistic consideration of the myriad of interacting issues associated in the original decision making process. Because of this single-mindedness, the resultant outcomes are often inefficient and develop unintended consequences (good and bad) associated with issues not considered when the decision is made. As John Muir's insights allude, decisions have ripple effects. For example, siting a new school in a particular location because it is the least expensive (sole criterion is economic) might miss the unintended consequences of higher fuel costs for longer bus routes, children's air pollutant exposures due to long bus rides or proximity to an interstate highway, or the social consequences of moving disadvantaged populations long distances out of their neighborhoods. Hence, the decision to place the school at a particular location based solely on short term economic criteria might not yield the best long term outcome for either the community's school budget or the children's welfare [3].

"Sustainability", the increasingly-discussed paradigm, does not refer to the current "sustainababble" frequenting many political and social discussions [4] rampant in the development of sustainable products, ideas and concepts—from "green" cleaning supplies to sustainable music, candy or sidewalks. Rather, the basic concept of sustainability is that put forward by the Brundtland Commission [5] which states that sustainable development is, "development that meets the needs of the present without compromising the ability of future generations to meet their own needs." This mirrors the policy of the Federal Government stated in the 1969 National Environmental Protection Act, which is "to create and maintain conditions under which humans and nature can exist in productive harmony, [and] that permit fulfilling the social, economic and other requirements of present and future generations"—the definition that the President used in Executive Order 13514 on sustainability and the federal government. In 2010, EPA provided an operational definition of sustainability: "Sustainability is the continued protection of human health and the environment while fostering economic prosperity and societal well being" [6].

In its report to the U.S. Environmental Protection Agency (EPA), the National Research Council recommended that "EPA formally adopt as its sustainability paradigm the widely used "three pillars" approach, which means considering the environmental, social, and economic impacts of an action or decision" [7] and furthermore that "EPA should also articulate its vision for sustainability and develop a set of sustainability principles that would underlie all agency policies and programs" [7]. EPA's Office of Research and Development (ORD) has developed research programs to support this sustainability paradigm [8], one of which is the Sustainable and Healthy Communities (SHC) Research Program. The SHC program fully embraces the "three pillars" approach described by the National Research Council and is developing tools, methods and approaches to support decisions that will foster community sustainability. Alternatively, there is a recent view [9,10,11,12] that, rather than examining sustainability simply as trade-offs among the three pillars, incorporates a broad range of criteria and objectives, drawing from the socio-ecological resilience literature and highlighting the importance of things like adaptive capacity and precaution in decision-making. These alternative approaches tend to emphasize the importance of taking paths that maximize gains (or avoid losses) in relation to the full range of sustainability criteria and have explicit rules for dealing with trade-offs.

11.2 PROGRAM HISTORY

Increasingly, reports appear on the unintended, usually negative, consequences of a legislative, policy or programmatic action, at neighborhood, community, city, county, state and national scales. For example, actions intended to increase energy supplies can have huge ramifications in the economic, environmental and social spheres and a decision targeted only at increased production can have significant unintended environmental or social consequences [13,14]. Similarly, a decision to site a waste management facility at a particular location solely based on economic criteria can result in major environmental justice issues, as these are often areas with disadvantaged populations [15,16]. Lacking a framework for decision making that includes consideration of the three pillars of sustainability—economics, environment, and social drivers—in an integrated and holistic fashion creates a high likelihood of unintended consequences.

In 2010, the Environmental Protection Agency's Office of Research and Development condensed its 13 topic-oriented research programs to focus on problems of broad national interest under principles of sustainability and solution-orientation. To that end, the AA realigned ORD into six programs focused on sustainability with regard to water, air (and climate/energy), chemicals, communities, human health risk assessment and homeland security (Figure 1).

The Sustainable and Healthy Communities Research Program (SHC) is shown in the figure as encompassing the entire research program because in order to assist communities holistically, SHC must avail itself of the results of the other five research programs. Of the six programs, the Sustainable and Healthy Communities Research Program (SHC) is arguably the most novel, adding a relatively new topic to, and audience for, ORD's research portfolio, and integrating the previous research programs (from the 13 topic-oriented) addressing ecosystem services, human health, geographic information systems, waste management, decision support and community engagement.

ORD had conducted community research in the past, but such work addressed specific issues in specific places. For SHC to address the new problems of broad national interest, simply "ramping up" site-specific research (more sites, more issues) was neither adequate nor feasible. Instead, EPA needed to define how to support the community decision-making process to advance their sustainability goals.

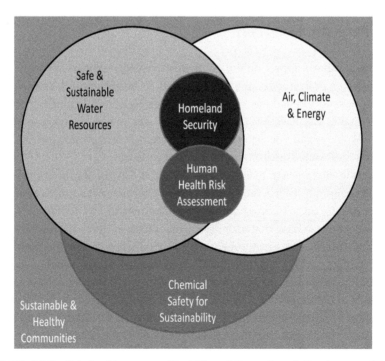

FIGURE 11.1 Relationships among Six Office of Research and Development (ORD) Sustainability Research Programs.

As such, the SHC Research Program is expressly focused on the growing interest of U.S. communities in sustainable practices [17]. In many ways, local communities are ahead of the sustainability curve, evidenced by participation in organizations that support communities' sustainability action. ICLEI (Local Governments for Sustainability) has 450 U.S. members, the U.S. Mayors Climate Action Agreement has 1060 signatories, and the Urban Sustainability Directors Network has over 100 actively-participatory members, while the American Planning Association has a Sustainability Division, the Center for Neighborhood Technology works on solutions and the Urban Land Institute provides economic perspectives for sustainability-related land use issues. What these organizations lack is the science for better evaluating problems and potential solutions, especially for aspects of human health, ecosystem services and environmental justice; SHC is designed to help provide this science basis.

While each community is unique, they have problems and decision issues in common, and as such, SHC needs to provide information and approaches that can be both flexible (i.e., can address different problems) and accessible to communities of varying size and scope (i.e., applicable at multiple spatial scales). In order to organize this "new approach", SHC engaged in a significant amount of outreach targeting new audiences (e.g., state and federal transportation agencies and regional planning groups) and a multiplicity of community types (e.g., small and large, rural, suburban and urban, agriculturally-based and manufacturing based) to ensure that SHC research projects would generate products that would be both useful and useable. The primary intent of this outreach effort— primarily listening sessions conducted in selected communities throughout the United States—was to determine what our "customers" needed, how they could use the information, and how this information could be used to overcome obstacles to decisions that advance sustainability goals. The most common needs expressed across communities was to create tools, methods and approaches to allow the holistic evaluation of community decision alternatives, and for metrics, indicators and indices to set sustainability goals and evaluate their progress. These priorities provided the architectural context of the Sustainable and Healthy Communities program.

11.3 PROGRAM GOALS AND OBJECTIVES

From its inception in 1970, EPA's mission has been to "protect human health and safeguard the environment—air, water and land—upon which life depends," and

ORD's pioneering environmental research has provided a sound science foundation for EPA's work. However, despite the successes of U.S. environmental legislation, and EPA policies and regulation, current trends in population, as well as in the production and use of energy, food, and materials, have strained our natural resource base and compromised the resilience of the environment. There are too many examples where human health and essential ecosystem functions have been negatively affected by cumulative exposures to multiple toxic pollutants and a changing physical environment. These impacts have economic costs (e.g., increased heating and cooling loads, costly burdens in infrastructure and municipal services, contamination of fisheries, and diminished access to clean drinking water) and societal costs (e.g., health impacts, disparities in health risks, and loss of natural areas for healthful recreation).

Community decisions that do not take into account the ripple effects end up with social, economic, and environmental trade-offs that are often not recognized, much less understood and considered. SHC, through its research and application of that research, will inform and empower those decision-makers affecting communities (including at federal, state and tribal levels) to effectively and equitably weigh and integrate human health, socio-economic, environmental, and ecological factors into their decisions in a way that those decisions better foster community sustainability. In particular, SHC seeks to provide information that will assist decision-makers in implementing innovative actions within communities and tribal programs that can complement EPA, state and tribal authorities and achieve shared sustainability goals in more flexible, economically beneficial and effectively synergistic ways. To put it in economic terms, we want to help communities make decisions that maximize positive externalities, while minimizing or eliminating negative externalities.

11.4   PROGRAM DESIGN

Each of ORD's programs has specific focal areas while maintaining close interrelationship with relevant parts of the other programs. Figure 1 illustrates the relationships among ORD's six sustainability research programs. These programs are using their expertise and experience to conduct transdisciplinary research which focuses on solving complex, real-world issues. They are seizing collaborative opportunities such that the relationships between them—safer chemicals and safer water supplies, less energy use, less air pollution and less waste—have implications for communities.

SHC is the focal point for ORD research on community sustainability, using systems-thinking, integrating perspectives from all realms (i.e., private, public and civil). Furthermore, the program emphasizes collaboration in order to transcend narrow boundaries of traditional disciplines and create new knowledge and new theory, fostering new practical applications that yield advantages for multiple beneficiaries. Functionally, this breadth means that SHC is also the focal point for research to support cross-cutting topics on children's health, community, health, and environmental justice, and will be developing ways to integrate research findings from all ORD programs. In Figure 1, this role is illustrated by SHC's location as the program that frames and encompasses the other five programs.

In order to accomplish this, SHC is organized into four themes. First, there is work on data and tools to support decision making. Second is research to characterize health and ecosystem linkages and impacts, to feed into decision support tools. Third is work that meets short term needs in communities with respect to legacy waste issues, like contaminated sites. However, this short term work also yields knowledge of processes that also feeds community decision support tools. Lastly, all these efforts are integrated into approaches for communities to use to evaluate decisions and find optimal solutions, including TRIO (Total Resources Impacts and Outcomes). The work of these four themes is described below.

11.5 DATA AND TOOLS TO SUPPORT SUSTAINABLE COMMUNITY DECISIONS

SHC is using decision science, interactive social media and sustainability assessment methods to assist communities in framing their sustainability goals and to develop new tools, indicators and spatial analyses for community use. In order to accomplish these activities, SHC scientists are working collaboratively with communities to develop ways to make data, information, and tools more interactive and more accessible to local audiences. Similarly, these scientists are compiling assessment indicators and tools and critiquing them for their applicability to community issues. This allows the creation of consistent metrics to characterize and communicate linkages among human health, well-being and environmental changes and measure progress toward sustainability goals. Three examples of research activities in SHC help to illustrate these approaches: (1) a classification of U.S. community types; (2) a national atlas for sustainability; and (3) an index of human well-being, described below.

(1) The statistical classification of U.S. communities will be used to guide development of decision and assessment tools that can address widely-shared sustainability issues. It will also inform transferability of tools to specific types of communities. The initial classification will be based on characteristics related to biophysical setting (e.g., climate, landform, soils, vegetation), community attributes (local governance, sustainability practices), demographic attributes (e.g., size, growth/decline, density, distribution) and ecosystem service characteristics. The classification will be updated over time to incorporate new data and relevant findings.

(2) The EnviroAtlas, a national Geographic Information System (GIS) atlas of sustainability-related parameters, will provide communities across the country with a suite of accessible, interactive maps showing indicators of production, demand and drivers of ecosystem services [18,19]. Categories of ecosystem services include: clean water for drinking; clean water for recreation and aquatic habitats; adequate water supply; food, fuel, and fiber; recreation, cultural and aesthetic amenities; contributions to climate stability; protection from hazardous weather; habitat and the maintenance of biodiversity; and clean air. A growing number of selected cities will have finer scale information with even more metrics.

(3) An index of human well-being [20,21,22] that would be applicable across spatial scales (national, regional, state, city, community, neighborhood) and temporal scales (intergenerational) is being developed by SHC. This index is comprised of information describing eight dimensions (health, safety and security, living standards, education, connection to nature, social cohesion, leisure time, and spiritual and cultural fulfillment) with each dimension having multiple indicators represented by multiple specific metrics. The index is being constructed to provide communities with a tool to assess the effects of decision options on the well-being of their residents, as well as those in adjacent and even distant communities. Obviously, development of these types of indicators and indices are challenging and often dependent of the specific value systems of individual communities or sensitive population groups (e.g., children, tribes, socio-economic entities).

11.6   FORECASTING AND ASSESSING ECOLOGICAL AND COMMUNITY HEALTH

SHC scientists conducting research in this area will develop the information and methods that communities need to assess the health and well-being of their residents. To accomplish this, these researchers conduct foundational research in two major areas: (1) the science of ecosystem goods and services—those ecosystem functions that society depends upon to survive and prosper—including their production, use and benefits; and, (2) the science of human health and well-being as influenced by changes in ecosystem services as well as exposures to chemicals and other stressors in homes, schools, or neighborhoods. SHC's ecosystem-focused research will develop methods to quantify ecosystem goods and services, such as, water filtration, nutrient recycling, and mitigation of floods and storm surges. The research addresses how to estimate current production of ecosystem goods and services, given the type and condition of ecosystems; how ecosystem services contribute to human health and well-being; and the way in which the production and benefits associated with ecosystem services may be affected under alternative decision scenarios, in order to address the sustainability of those functions.

SHC's human health-focused research: will develop better methods to quantify, track, and reduce cumulative risks to public health; will develop a holistic understanding of how children's health may be linked to exposures (from before birth through adolescence) and impact their health throughout life; and will understand how differences in community setting (e.g., location of residence in relative to pollution sources, availability of safe, walkable streets, access to healthy foods) can contribute to good health and well-being or to environmental injustice and disproportionate health risks. Communities can use these types of information to develop and implement better public health policies and practices, especially for their most vulnerable residents (children, the elderly, or socio-economically disadvantaged), and to evaluate the effectiveness of interventions designed to improve public health.

Some examples of primary research in this area of SHC are: (1) Methods to Enhance Children's Health; (2) Standardized Classification of Ecosystem Goods and Services (EGS); (3) Searchable Database of Ecosystem Services; and (4) Web-based Tools for Environmental Justice, which are described below.

(1) Children's health research will contribute to EPA risk assessments, guidance documents and policies that protect overall children's health by providing metrics for age-specific chemical and non-chemical exposure and health impacts. In addition, work will examine children's health in a holistic way, looking at a wide variety of factors (e.g., children's play, psycho-social issues, their surrounding built and natural environments) and how they may interact with chemical and non-chemical exposures to impact children's health and health disparities [23,24].

(2) A central scientific problem limiting the clear understanding and consistent linkage of ecosystem changes to human health and well-being is having a metric with which to compare functions across different geographic settings—e.g., an acre of wetland in one location will not contain the same kinds and amounts of natural functions as an acre of wetland elsewhere. For EGS classification, SHC will develop standardized metrics for ecosystem goods and services; thus, significantly enhancing evaluation of how policy choices affect human health and well-being conditions. In addition, it will allow "trading" of ecosystem service credits, informing more commensurate mitigation of ecosystem damages through a consistent quantification of services that were lost.

(3) SHC researchers are developing production functions for many U.S. ecosystem services and benefits, that is, a characterization of the kind and amount of services and benefits a given unit of each ecosystem will produce. This is being accomplished by developing protocols for estimating the value of ecosystem services, including methods for quantifying the uncertainty associated with these estimates, understanding how scale affects estimates, and assessing the transferability of results from one area to other areas. These production functions are being catalogued and will be easily accessible to EPA, other agencies, NGOs, and anyone interested in considering ecosystem service trade-offs associated with changes in environmental conditions or decision alternatives.

(4) SHC is developing user-friendly web-based tools to help communities assess whether disproportionate health impacts or environmental exposures exist and, if so, to develop risk mitigation strategies that advance environmental justice. With this type of process and substance assistance (e.g., defining objectives, creating partnership databases, ranking risks and developing mitigation options), communities can better locate the source of the problems and improve conditions for everyone.

11.7   IMPLEMENTING NEAR-TERM APPROACHES FOR SUSTAINABLE SOLUTIONS

Research in this area of SHC builds upon federal, regional and state successes and experience to improve the efficacy of methods and guidance to address existing sources of land and groundwater contamination (required under RCRA [25] and Superfund [26]). RCRA, which regulates land-based disposal of waste (and focuses on hazardous waste) has the goal of reducing waste and encouraging recycling. This is not a ban on land-based disposal, but rather a regulation thereof, which uses "manifests" and the "cradle-to-grave" tracking system. All hazardous waste must obtain an identification number, and be accompanied by a "manifest" which tracks the waste. Each time the waste changes hands, a copy is sent back, ensuring that everyone along the chain is informed, and preventing unidentified wastes from arriving at disposal facilities. Superfund is a United States federal law designed to clean up sites contaminated with hazardous substances. Superfund created the Agency for Toxic Substances and Disease Registry (ATSDR), and it provides broad federal authority to clean up releases or threatened releases of hazardous substances that may endanger public health or the environment. The law authorized the Environmental Protection Agency (EPA) to identify parties responsible for contamination of sites and compel the parties to clean up the sites. Where responsible parties cannot be found, the Agency is authorized to clean up sites itself, using a special trust fund. SHC research builds on RCRA and Superfund policies that encourage the use of innovative approaches to reduce new sources of contamination; enable material and energy recovery from existing waste streams; and enable brownfields sites to be put to new, economically productive uses that benefit communities. SHC research to address near-term solutions includes management of contaminated sites, materials and waste management, integrated management of reactive nitrogen, EPA's Report on the Environment, and sustainable technologies. Specific examples of research in this area are (1) tools to assess, measure and monitor clean-up of contaminated sediments; (2) beneficial reuse of material and energy recovery from wastes; and (3) sustainable nitrogen management.

(1) SHC research will improve biological, chemical and geophysical procedures to assess chemicals in sediments [27,28,29,30,31], as well as to better predict chemical concentrations in fish, shellfish, and birds (i.e., aquatic dependent wildlife) from exposure to contaminated sediments.

These will allow communities to measure and document the effectiveness of sediment remediation.

(2) Beneficial reuse research will provide data and tools to help optimize the recovery of energy from wastes and the beneficial reuse of wastes [32,33], thereby identifying opportunities to further reduce the volume of waste disposed, conserve natural materials and reduce net costs while protecting the natural environment in an economically and technically sound manner.

(3) When reactive nitrogen is released to the environment it creates a cascade of harmful effects that includes eutrophication of aquatic ecosystems, toxic algal blooms [34], hypoxia or "dead zones" [35,36,37], acid rain, nitrogen saturation of forests, contributions to global warming, and human health effects due to contamination of drinking water and air pollution [38]. SHC nitrogen research is part of an agency-wide effort. This work will synthesize existing and new analyses about the sources of nitrogen, its distribution in air, land and water, and its impacts on valuable ecosystem services [39], then it will identify strategic and efficient options to reduce the most damaging effects of reactive nitrogen while maintaining the benefits of nitrogen use.

11.8   INTEGRATED SOLUTIONS FOR SUSTAINABLE OUTCOMES

The two primary significant barriers to effective decision making for community sustainability are the failure or inability to account for unintended impacts of actions and the failure to account for and take advantage of linkages among issues [40,41,42,43]. Regardless of the reason (oversight or lack of information), these omissions impede effective decision making. The design of policies, technologies and incentives to best foster community sustainability needs to take into account the linkages between human health and welfare and the built and natural environments, especially with respect to ecosystem services. SHC research in this area is exploring systems modeling approaches to account for the linkages among resources and assets managed by a community with special emphasis on high priority decision sectors—waste and materials management, land use, transportation, and buildings and infrastructure. Systems models that account for stocks and flows of energy, materials and water can be used by communities to identify opportunities to increase efficiencies and resource recovery with their actions.

TRIO (Total Resource Impacts and Outcomes—a term coined by SHC) research is developing methods and approaches to account for the multiple and interconnecting implications of decision alternatives, including direct and indirect costs and benefits across dimensions of human and community well-being (economic, environmental and societal). A transdisciplinary team of health scientists, ecologists, economists and policy partners will evaluate and develop indicators that reflect the response of these sustainability dimensions to decisions made by communities. The TRIO approach will use systems models to estimate the full range of costs, benefits, impacts and outcomes for a given decision using relative weights for specific indicators to reflect community preferences and needs. TRIO is being tested in a proof-of-concept project in Durham, NC, but ultimately the TRIO tool will be available as a web-based model for more widespread application to community sustainability decisions.

11.9   SUMMARY AND CONCLUSIONS

Each community is unique with respect to policy context, resources, constraints and culture, but the issues of sustainability are common to all—a clean environment, a robust, resilient economy and concerns for their residents' health and well-being. The desired goal of the Sustainable and Healthy Communities research program is to provide communities the information they need to transform their expressed interest in sustainability into integrated actions that can address their short term needs, but yield greater benefits than current piecemeal approaches—a laudable goal for a program in its developmental stage. To accomplish this goal, SHC will develop and use a whole-systems approach to proactively and holistically assess the implications of community-level decisions, identify negative unintended consequences and evaluate opportunities for achieving optimized outcomes through integrated sustainability practices.

The tools and methods developed by SHC will enable EPA, regions, states, tribes and communities to implement their respective responsibilities with far greater ability to capture synergies in meeting their respective sustainability goals. The information from the SHC research program, together with communities' more intimate connections with their place, as well as with local residents, businesses and other groups, provides opportunities for communities to pursue effective, state-of-the-art actions.

There is also great interest from communities, around the country and the world, in using more sustainable practices to provide a full range of services (economic, environmental and social) [40,41,42,43]. These conditions provide both a receptive audience for SHC research products and an expansive level of information about early experiences on which to build and refine a scientific research program with immediate applicability to community needs. Supported by tools and information developed by SHC, communities, individuals and organizations can be empowered to better understand and manage how their activities can promote progress toward a sustainable future. As benefits accrue for individual communities, and as lessons spread, more and more sustainable communities will add up to a more sustainable nation and planet.

REFERENCES

1. Bade, W.F. The Life and Letters of John Muir; Houghton Mifflin Company: Boston, MA, USA, 1924.
2. Clark, T.N. Comparative Study of Community Decision-Making; Inter-university Consortium for Political and Social Research: Ann Arbor, MI, USA, 1999.
3. McDonald, N.C. School siting: Contested visions of a community school. J. Am. Plann. Assoc. 2010, 76, 184–198.
4. The Worldwatch Institute. State of the World 2013: Is Sustainability Still Possible; Island Press: Washington, DC, USA, 2013.
5. World Commission on Environment and Development. Our Common Future; Oxford University Press: Oxford, UK, 1987.
6. World Commission on Environment and Development. Our Common Future; Oxford University Press: Oxford, UK, 1987.
7. National Research Council. Sustainability and the U.S. EPA; The National Academies Press: Washington, DC, USA, 2011.
8. U.S. Environmental Protection Agency. True North: Sustainability Research at EPA. Available online: http://www.epa.gov/sciencematters/april2011/truenorth.htm (accessed on 2 December 2013).
9. Gibson, R.B.; Hassan, S.; Holtz, S.; Tansey, J.; Whitelaw, G. Sustainability Assessment: Criteria and Processes; Earthscan: London, UK, 2005.
10. Kemp, R.; Parto, S.; Gibson, R.B. Governance for sustainable development: Moving from theory to practice. Int. J. Sustain. Dev. 2005, 8, 12–30.
11. Gibson, R.B. Beyond the pillars: Sustainability assessment as a framework for effective integration of social, economic and ecological considerations in significant decision-making. J. Environ. Assess. Policy Manag. 2006, 8, 259–280.
12. Gibson, R.B. Sustainability assessment: Basic components of practical approach. Impact Assess. Proj. Apprais. 2006, 24, 170–182.
13. Stiglitz, J.E.; Sen, A.; Fitoussi, J.-P. Report by Commission on the Measurement of Economic Performance and Social Progress. Available online: http://www.stiglitz-sen-fitoussi.fr/documents/rapport_anglais.pdf (accessed on 2 December 2013).
14. Gordon, R. Climate change and the poorest nations: Further reflections on global inequality. Univ. Colo. Law Rev. 2008, 78, 1559–1624.

15. Toxic Wastes and Race in the United States: A National Report on the Racial and Socio-Economic Characteristics with Hazardous Waste Sites. Available online: http://www.ucc.org/about-us/archives/pdfs/toxwrace87.pdf (accessed on 2 December 2013).
16. Toxic Wastes and Race at Twenty 1987–2007. Available online: http://www.ucc.org/justice/pdfs/toxic20.pdf (accessed on 2 December 2013).
17. ICMA 2010 Sustainability Survey Results. Available online: http://icma.org/en/icma/knowledge_network/documents/kn/Document/301646/ICMA_2010_Sustainbility_Survey_Results (accessed on 2 December 2013).
18. EnviroAtlas. Available online: http://www.epa.gov/research/enviroatlas/ (accessed on 2 December 2013).
19. Developing the "EnviroAtlas" to Support Community Decisions. Available online: http://www.epa.gov/research/annualreport/2012/enviroatlas.htm (accessed on 2 December 2013).
20. Summers, J.K.; Smith, L.M.; Case, J.L.; Linthurst, R.A. A review of the elements of human well-being with an emphasis on the contribution of ecosystem services. Ambio 2012, 41, 327–340. [Google Scholar] [CrossRef]
21. Smith, L.M.; Case, J.L.; Smith, H.M.; Harwell, L.C.; Summers, J.K. Relating ecosystem services to domains of human well-being: Foundation for a U.S. index. Ecol. Indic. 2013, 26, 79–90.
22. Smith, L.M.; Case, J.L.; Harwell, L.C.; Smith, H.M.; Summers, J.K. Methods for developing relative importance values: Assessing relationships between ecosystem services and elements of human well-being. Hum. Ecol. 2013.
23. HIA Case Studies. Available online: http://www.epa.gov/research/healthscience/hia-case-studies.htm (accessed on 2 December 2013).
24. EPA/NIEHS Children's Environmental Health and Disease Prevention Research Centers (CEHCs). Available online: http://epa.gov/ncer/childrenscenters/ (accessed on 2 December 2013).
25. RCRA—Resource Conservation and Recovery Act of 1976. Non-hazardous Waste Regulations. Available online: http://www.epa.gov/osw/laws-regs/regs-non-haz.htm (accessed on 2 December 2013).
26. Superfund: Superfund is the common name for the Comprehensive Environmental Response, Compensation, and Liability Act of 1980. Superfund Regulations & Enforcement. Available online: http://www.epa.gov/superfund/policy/remedy/sfremedy/regenfor.htm (accessed on 2 December 2013).
27. Southerland, E.; Kravitz, M.; Wall, T. EPA's Contaminated Sediment Management Strategy; Lewis Publishers: Boca Raton, FL, USA, 1992.
28. McCauley, D.J.; DeGraeve, G.M.; Linton, T.K. Sediment quality guidelines and assessment: Overview and research needs. Environ. Sci. Policy 2000, 3, 133–144.
29. Tolaymat, T.; U.S. Environmental Protection Agency. Monitoring Approaches for Landfill Bioreactors; National Risk Managment Research Laboratory, Office of Research and Development, US Environmental Protection Agency: Cincinnati, OH, USA, 2004.
30. Nelson, W.G.; Bergen, B.J. The New Bedford Harbor Superfund site long-term monitoring program (1993–2009). Environ. Monit. Assess. 2012, 184, 7531–7550.
31. Burkhard, L.P.; Mount, D.R.; Highland, T.L.; Hockett, J.R.; Norberg-King, T.; Billa, N.; Hawthorne, S.; Miller, D.J.; Grabanski, C.B. Evaluation of PCB bioaccumulation by Lumbriculus variegatus in field-collected sediments. Environ. Toxicol. Chem. 2013, 32, 1495–1503.

32. Overcash, M.; Sims, R.C.; Sims, J.L.; Nieman, J.K. Beneficial reuse and sustainability: The fate of organic compounds in land-applied waste. J. Environ. Qual. 2005, 34, 29–41.

33. Innovative Waste Consulting Services. Data Gap Analysis and Damage Case Studies: Risk Analyses from Construction and Demolition Debris Landfills and Recycling Facilities. Report to U.S. Environmental Protection Agency; Report Number 0212041.003.030; U.S. EPA: Cincinnati, Ohio, USA, 2012.

34. Chesapeake Bay Foundation. 2012 State of the Bay Report; Chesapeake Bay Foundation: Washington, DC, USA, 2012; p. 20. Available online: http://www.cbf.org/about-the-bay/state-of-the-bay/2012-report (accessed on 2 December 2013).

35. Diaz, R.J.; Rosenburg, R. Spreading dead zones and consequences for marine ecosystems. Science 2008, 321, 926–929.

36. Turner, R.E.; Rabalais, N.N.; Justic, D. Gulf of Mexico hypoxia: Alternate states and a legacy. Environ. Sci. Technol. 2008, 42, 2323–2327.

37. Greene, R.M.; Lehrter, J.C.; Hagy, J.D. Multiple regression models for hindcasting and forecasting midsummer hypoxia in the Gulf of Mexico. Ecol. Appl. 2009, 19, 1161–1175.

38. U.S. Environmental Protection Agency (EPA). Reactive Nitrogen in the United States: An Analysis of Inputs, Flows, Consequences, and Management Options; A Report of the EPA Science Advisory Board; U.S. EPA, 2011.

39. Compton, J.E.; Harrison, J.A.; Dennis, R.L.; Greaver, T.L.; Hill, B.H.; Jordan, S.J.; Walker, H.; Campbell, H.V. Ecosystem services altered by human changes in the nitrogen cycle: A new perspective for US decision making. Ecol. Lett. 2011, 14, 804–815.

40. Ceres, Mobilizing Business Leadership for a Sustainable World. Available online: http://www.ceres.org/ (accessed on 2 December 2013).

41. State of Green Business Report 2013. Available online: http://www.greenbiz.com/research/report/2013/02/state-green-business-report-2013 (accessed on 2 December 2013).

42. SustainableBusiness.com. Available online: http://www.sustainablebusiness.com/ (accessed on 2 December 2013).

43. U.S. Green Building Council. Available online: http://usgbc.com/ (accessed on 2 December 2013).

Keywords

- Health inequities
- Deprivation
- Glasgow
- post-industrial
- vacant and derelict land
- brownfields
- PARDLI index
- open space
- greenspace
- urban agriculture
- community gardens
- community regeneration
- urban planning
- Healthy communities
- Physical activity
- Built environment
- Parks
- Trails
- Land use
- Urban design
- Schools
- Workplace
- Transportation
- Urban ecology
- stewardship
- environmental planning
- Vacant land
- decision support
- community perspectives

- land cover change
- urbanization
- scenarios
- stakeholder participation
- GIS
- port city
- urban landscape
- multifunctional landscape
- stakeholder-based model
- backcasting
- forecasting
- sustainable development
- neighbourhood regeneration
- community land trusts
- housing commons
- guerrilla gardening
- sustainability
- decision making
- community
- TRIO

Author Notes

Chapter 1

Acknowledgements

This research is part of a larger project involving a comparison of Glasgow and New York City regarding the relationship between environmental health justice and the built environment. I conducted most of the research reported in this paper while a Fulbright Distinguished Chair at the Glasgow Urban Laboratory, a collaboration of the Mackintosh School of Architecture(MSA)/Glasgow School of Art, and the Glasgow City Council. I would like to thank everyone connected with those institutions, but especially Julia Radcliffe and Brian Evans of the Glasgow Urban Lab/MSA; Alistair MacDonald, Gerry Grams, Jamie Arnott, and Cathy Johnston of the Glasgow City Council Development and Regeneration Services (GCC DRS), and Jim Gordon and Fiona Hunter from the DRS GIS unit. There were many other people who were helpful during my 6 months in Glasgow, including Russell Jones, David Walsh, and the entire staff at the Glasgow Centre for Population Health; Anne Ellaway, of the U.K. Medical Research Council; Alastair Corbett, of the Glasgow and Clyde Valley Green Network Partnership; Jamie Pearce, of the Centre for Research on Environment, Society, and Health, the School of GeoSciences at the University of Edinburgh; and many of the faculty at the University of Glasgow, Strathclyde University in Glasgow, and of course, all of my colleagues (and the students) at the Mackintosh School of Architecture/Glasgow School of Art. I would also like to thank my colleagues at my home institution of City University of New York, especially Andrew Maroko, Stefan Becker, and Brian Morgan, who "held down the fort" with the GISc Program and the Urban GISc Lab, and taught my courses whilst I was in Glasgow for six months. My deep appreciation and admiration also goes to my CUNY doctoral student, Gretchen Culp, cartographer extraordinaire, for translating my working maps into publication-quality art. Most of all, I would like to acknowledge the US-UK Fulbright Commission, and the efforts of their London-based staff, who gave me this unique opportunity to immerse myself in Glaswegian culture and environment, which provided me with access to the human and data resources that were necessary for this study to take place.

Chapter 2

Competing Interests

The authors declare the following potential competing financial interests. Sallis: Nike, Inc., director of grant that supported preparation of this paper. Santech, Inc., shareholder. No relation to current paper. Oregon chapter of the American Planning Association, consultant. No conflict. Thai Health Promotion Foundation, consultant and speaker. No conflict. Cavill, Gebel, Parker, Ding: Nike, Inc., paid consultants for research work done in preparation of this paper. Spoon, Engelberg, Lou, Thornton, Wilson, Cutter: no conflict.

Authors' Contributions

JS conceived of the study and its design, synthesized and interpreted data and drafted the manuscript. CS managed and participated in literature reviews, synthesized data and helped draft the manuscript. DD participated in literature reviews, synthesized data and helped draft the manuscript. All remaining authors completed literature reviews and synthesized data. All authors read and approved the final manuscript.

Acknowledgments

This project was supported by Nike, Inc and Active Living Research, a program of the Robert Wood Johnson Foundation.

A longer version of the present study, with additional sections, is posted on the Active Living Research website (http://activelivingresearch.org/making-case-designing-active-cities), along with the data tables with codes for each finding.

Thanks to the content experts from multiple disciplines and sectors who provided input on the review process: *Open Spaces/Parks/Trails*: Ariane Rung, Louisiana State University, Andrew Mowen, The Pennsylvania State University, Zarnaaz Bashir, National Recreation and Park Association, Karla Henderson, North Carolina State University; *Urban Design/Land Use*: Anne Vernez Moudon, University of Washington, Andrew Dannenberg, University of Washington, Reid Ewing, University of Utah; *Transportation*: Robert Cervero, University of California, Berkeley, Charlie Zegeer, University of North Carolina Highway Safety Research Center, Chris Kochtitzky, Centers for Disease Control and Prevention; *Schools*: Jeff Vincent, Center for Cities and Schools, Nisha Botchwey, Georgia Institute of Technology; *Workplaces/Buildings*: Gayle Nicoll, Ontario College of Art & Design.

Chapter 3

Author Contributions

Wrote the first draft of the manuscript: JC. Contributed to the writing of the manuscript: JC AC.ICMJE criteria for authorship read and met: JC AC. Agree with manuscript results and conclusions: JC AC. Worked in Richmond: JC AC. Worked in Nairobi: JC.

Chapter 5

Acknowledgements

The authors would like to acknowledge the critical local research participation of the Urban Ecology Institute (UEI), Urban Resources Initiative(URI), Nine Mile Run Watershed Association, NYU/Wallerstein Collaborative for Environmental Education, Parks & People Foundation, Casey Tree Endowment Fund, the Steering Committee of the Urban Ecology Collaborative and finally, the USDA Forest Service Northeastern Area State and Private Forestry for multi-city research support and collaboration.

Chapter 6

Conflict of Interests

The author declares that there is no conflict of interests regarding the publication of this paper.

Chapter 8

Acknowledgments

This research was supported by the U.S. National Science Foundation, grant #1226629, and by the Institute for Sustainable Solutions at Portland State University. We would like to thank John Barnes of the City of McMinnville Planning Department, Bob Harmon of the Oregon Department of Water Resources, Lesley Hegewald of the Mid-Willamette Valley Council of City Governments, Sarah Marvin of the Oregon Department of Land Conservation and Development and Jan Wolf of the City of Newberg Engineering Department for providing GIS data for our study area. We would like to thank the members of the workshop for their invaluable input and for their helpful comments. The views expressed are our own and do not necessarily reflect those of sponsoring agencies.

Author Contributions

R.W.H. and H.C. conceived and designed the presented research. R.W.H. and H.C. both consulted with stakeholder participants. R.W.H. performed the research and analyzed the data. R.W.H. and H.C. wrote the paper.

Conflicts of Interest

The authors declare no conflict of interest.

Chapter 9

Acknowledgments

The authors wish to thank several Romanian students, who as part of their internship assisted in collecting the necessary data.

Conflicts of Interest

The authors declare no conflict of interest.

Chapter 10

Acknowledgements

I would like to thank the following for their critical feedback and encouragement: Melissa García Lamarca, Abby Gilbert, Graham Haughton, Stephen Hincks, Philipp Horn, Melanie Lombard, Andy Merrifield, Michele Vianello, and all at the Urban Rights Group at Manchester. This work was supported by the ESRC (ES/J500094/1).

Chapter 11

Acknowledgements

The authors would acknowledge the contributions of Kathryn Saterson, Rochelle Araujo, Iris Goodman and Rick Linthurst to the early development of the Sustainable and Healthy Communities Research Program (SHC). In addition, we acknowledge the continuing contributions of following members of the SHC Team—Kathryn Saterson, Valerie Zartarian, David Kryak, Randy Parker, and Abdel Kadry.

Disclaimer

The information in this document has been funded wholly (or in part) by the U.S. Environmental Protection Agency. It has been subjected to review by the National Health and Environmental Effects Research Laboratory and approved for publication. Approval does not signify that the contents reflect the views of the Agency, nor does mention of trade names or commercial products constitute endorsement or recommendation for use.

Conflicts of Interest
The authors declare no conflict of interest.

Index